Quarto.com

© 2024 Quarto Publishing Group USA Inc.
Text © 2024 Pete Evanow
Photography © Ford Motor Company images used with permission, unless
otherwise noted

First Published in 2024 by Motorbooks, an imprint of The Quarto Group,
100 Cummings Center, Suite 265-D, Beverly, MA 01915, USA.
T (978) 282-9590 F (978) 283-2742

Motorbooks titles are also available at discount for retail, wholesale, promotional,
and bulk purchase. For details, contact the Special Sales Manager by email at
specialsales@quarto.com or by mail at The Quarto Group, Attn: Special Sales
Manager, 100 Cummings Center, Suite 265-D, Beverly, MA 01915, USA.

28 27 26 25 24   1 2 3 4 5

ISBN: 978-0-7603-8333-9

Digital edition published in 2024
eISBN: 978-0-7603-8334-6

Library of Congress Cataloging-in-Publication Data

Names: Evanow, Pete, 1957- author.
Title: Ford Bronco : the original SUV / Pete Evanow.
Description: Beverly, MA : Motorbooks, an imprint of The Quarto Group,
    2023. | Includes bibliographical references and index. | Summary:
    "Ford Bronco offers a complete history, from the original Bronco's
    introduction as a 1966 model through the following four generations
    ending in 1996 all the way to Ford's all-new, brilliantly styled Bronco
    introduced for the 2021 model year"— Provided by publisher.
Identifiers: LCCN 2023033459 | ISBN 9780760383339 |
    ISBN 9780760383346 (ebook)
Subjects: LCSH: Bronco truck—History. | Ford trucks—History. |
    Sport utility vehicles—History.
Classification: LCC TL230.5.B76 E93 2023 | DDC 629.222/2—dc23/eng/20230731
LC record available at https://lccn.loc.gov/2023033459

Design: Cindy Samargia Laun
Cover Image: 2022 Ford Bronco. Ron Kimball/KimballStock
Back Flap: Author Pete Evanow standing next to his restored 1965 Ford Ranchero.
    Courtesy of the author.
Page Layout: Cindy Samargia Laun

Printed in China

Ford

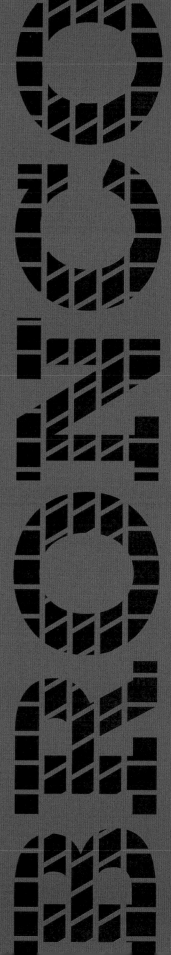

The Original SUV

PETE EVANOW

Foreword by Shelby Hall

CONTENTS

This book is dedicated to my old friend and colleague Kevin Kennedy, who devoted his incredible career to all things Blue Oval and made sure the right Ford logo was worn toward the camera, and the correct statement was issued at the precise moment. Congratulations on all your hard work and thank you for your friendship over the years.

And to Donald Frey. You were not only Time *magazine's "sharpest idea man," but clearly someone who led an amazing life, and as William White, Mr. Frey's successor at Bell & Howell, noted, "one of the greatest people you'd ever want to meet." Cheers to the man behind the Mustang and the Bronco.*

Finally, to all the people at the Michigan Assembly Plant. To the union members and management alike—everyone over the decades who has been responsible for building the Bronco, an icon in the automotive world. Now, to have the chance to extend that legacy, keep up the great work and Godspeed.

A True Family Legacy

My childhood memories begin with Baja: sagebrush, off-road trucks, and my family all around. As the youngest granddaughter of the legendary Rod Hall, eating dirt and watching desert sunsets was the norm. They knew I was made for the adventure life when I would beg my dad to "Go Faster!" while taking me for a spin in his race truck, all while not even being big enough to see over the dash.

My yearning for more dirt therapy came in 2015 during the restoration of my grandfather's famed 1968 Baja-winning Bronco. Of course, I knew its history, but in the years that followed, I learned its true narrative not only between it and my grandfather, but the story that Bronco and I would go on to create. This narrative was one of passion, sacrifice, tenacity, and teamwork.

In 2019, when Ford came to our hometown to thank my Grandfather for earmarking the Bronco as a true off-roading vehicle, they also invited me to be a driver for the Bronco R Baja race team. I felt like Cinderella! This was one of the world's most awaited vehicle launches, and we were taking the prototype to Baja—back to where it all began. Thus 2020 became a year of exceptional memories: through my own passion, sacrifice, tenacity, and teamwork, Bronco R crossed the Baja 1000 finish line, and as an all-female team, Penny Dale and I won the Rebelle Rally, bringing home the inaugural first place finish for the new generation Bronco family.

My personal Bronco has changed my life: from competing against the elite; empowering other Bronco-owning women to conquer their fears; learning to slow it down and enjoy the sights; to getting to see the United States through many wonderful road trips; and today, proudly holding the torch my grandfather passed down. I am honored to be a member of the Blue Oval Family and to represent the grassroots of off-road. I hope the Bronco narrative speaks to you, too, in your own unique way.

Here's to new dirt roads and always saying yes to the adventure!

—Shelby Hall

The famous Product Development Center on Oakwood
Avenue in Dearborn, Michigan, where for more than
30 years Ford's most brilliant designs were created,
including Bronco. © 2023 Ford Motor Company.

Chapter 1

BLUE OVAL ICONS
at Oakwood

Imagine the Ford Design studio in the early 1960s. The male-dominated staff assembled within the impressive mid-century building on Oakwood Avenue in Dearborn, Michigan, formally known as the Product Development Center (PDC), is hunkered down over their drawing tables creating some of the most successful, and future-legendary, vehicles the Blue Oval would ever produce: Mustang, Thunderbird, upgraded F-100 pickup, Galaxie 500, Lincoln Continental, GT40 (with the help of the High Performance and Special Models Operation Unit), and so many others.

However, there was one particularly unusual styling effort that would soon become a true stand-out design and yet another sales victory: the Bronco. In its first year alone, 1966, it would attract nearly 24,000 buyers, eventually tallying more than a quarter of a million vehicles throughout its first generation. It was an instant classic.

What made this possible?

THE PERFECT BLEND . . .

A combination of timing, superb design, a unique selling proposition, and good fortune ensured the Bronco's success. Buyers were looking for options beyond Jeep, seeking a short wheelbase, four-wheel drive "service" or "civilian" truck (in Ford's parlance, "off-road vehicle," a.k.a. ORV), and the Bronco provided the answer. Ford's marketing and sales teams, in association with Vice President for Styling Gene Bordinat's design studio, created the perfect vehicle for the moment. As America evolved into a more demanding, independent-minded society in the 1960s, people were searching for an unconventional form of transportation, for something that wasn't in their neighbor's driveway. The first generation two-door Bronco initially came in three body styles: wagon, half cab pickup, and open-body roadster, each with its own personality. Bronco's success became just as much about choice and freedom as it was about dynamic purpose.

(continued on page 12)

Built for the 1934 Chicago World's Fair, the Ford Rotunda was intended as a tourist attraction and was subsequently moved to Dearborn, Michigan, where it became the visitor's center for Ford's corporate headquarters. Its striking Albert Kahn design was a tremendous draw. Sadly it burned down in 1963. © *2023 Ford Motor Company.*

It's no secret that the 1960s was an incredible period for design. True, the previous decade had produced some classic vehicles, the rise of chrome and fins, and a gaggle of prototypes, but the 1960s was so different in every way—there was music from Elvis, the Beatles, the Rolling Stones, the coolness of the Rat Pack, and the promise of JFK. It was an exciting period, and designers and stylists, forever at the apex of popular culture, consumer taste, and the latest trends, introduced novel creative standards that became pillars of contemporary expression.

This was a moment when conservatism began to take a back seat to a growing liberation all reflected within the framework of every new vehicle to come out of Ford's Rotunda.

A talented pool worked within the Product Development Center (PDC) in Dearborn, Michigan, originally commissioned by Henry Ford II in May 1953 to house all design, research and development, and consumer analysis. With its wooden floors, mahogany wall panels, and private courtyard—highlighted by the centerpiece tribute to the original

Ford Rotunda—the center served the Blue Oval and its design staff for 67 years.

Armed with any number of new product assignments approved by the Product Planning Committee, as well as redesigns and upgrades to existing models, the PDC hummed mightily throughout the decade as some of Ford's key vehicles were created and positioned alongside future offerings—Mustang and Bronco being just two. Tucked away quietly was the development of the zenith of competition cars, the GT40.

Alongside all this action, the 1965 Galaxie debuted a new look, with vertically stacked dual headlights for the first time. Now taller and heavier, the revised vehicle rode on a new suspension system, moving from a leaf-spring rear to a three-link system featuring coil springs. By 1966 the Galaxie 500 Convertible would take third in the open-top sales category with 27,454 purchased, trailing Mustang's 72,119 convertibles that same model year.

In the truck arena, Ford's popular F-Series received a new mid-cycle look in 1965. It was a significant upgrade that included an innovative platform highlighted by the Twin I-Beam front suspension, a technology still in use today. The model name Ranger was revived—it had last been worn by a base model of the defunct Edsel lineup, but was now identified as a high-level styling package for the pickup line.

In 1964 the engineering constituent of the race program charged with shaming Ferrari in long-distance competition had moved to Dearborn from the United Kingdom to refine what would become the record-setting GT40. Two years later, the Shelby American–perfected Mk II version took the checkered flag at Le Mans and sent Ferrari packing.

It was a grand time for Ford.

After the Ford Rotunda was moved from its brief residence in Chicago to Dearborn, the structure was updated by its designer, Albert Kahn, with interior murals showing the River Rouge assembly line. The structure served as Ford's welcoming center—and new-model showcase—from 1936 until 1963, with the exception of World War II when its theater was used as a movie hall to entertain troops. Inventor R. Buckminster Fuller was responsible for designing the hallmark lightweight geodesic dome for which the Rotunda was best known. © 2023 Ford Motor Company.

(continued from page 9)

Naturally, competition existed in the marketplace. The original World War II veteran, the Jeep, which was initially sold as the Willys-Overland, became the "civilian Jeep" (CJ) and owned the lion's share of ORV sales. In 1961 came the International Harvester Scout, presented as a commercial utility pickup or, optionally, with a full-length rooftop. Toyota's Land Cruiser had limited sales in its early years in the United States, but was growing as Toyota's presence in the States expanded.

Interestingly Ford had been one of three auto manufacturers assigned to develop and build military jeeps for use during World War II, so the experience to create a short-wheelbase four-wheel drive vehicle was already resident in-house. Drawing on that experience, the new Bronco, originally the idea of Ford product virtuoso Donald N. Frey (see sidebar), became the hit Ford had hoped for. The Bronco's combination of off-road prowess and highway civility quickly earned the interest, respect, and endorsement of adventurers and outdoor enthusiasts throughout the country. It didn't hurt that legendary off-road racer Bill Stroppe introduced a team of modified Broncos to compete in numerous races including the Baja 1000. Given an early half cab by Ford, even before it was available to the public, Stroppe converted that first six-cylinder Bronco into a successful racer and overall winner at the 1967 Riverside Four-Wheel Drive Grand Prix. Ford promoted that accomplishment in its advertising, and customers, particularly those running businesses that would call on the new vehicle's off-road capabilities, were ready to put Bronco to the test.

No doubt, the 1960s were a good time to be in Ford's design studio. Catching lightning in a bottle seemed an almost common occurrence. Of course, this success wasn't just luck. It was a case of understanding the customer, the competition, and the importance of having what is known in marketing as a "unique selling proposition." With the Bronco sold on dealership floors alongside the Mustang, F-100, and other great models, Ford was in the driver's seat—and so were its customers.

CONCEPTUALIZING THE NEW OFF-ROAD VEHICLE

Who was a likely customer for the proposed Bronco? In 1963 Ford's designers were already at work on the Mustang, and they wanted to add something like a four-wheel drive sports car to the lineup. Playing off the ongoing discussion of horse symbolism, they eventually assigned the design the code name "Bronco." Marketers within Ford World Headquarters, also known as the "Glass House," also suggested "Wrangler", but obviously Ford did not trademark it. (Twenty years later, the moniker would become the new identity for Jeep's two-door.) Eventually the internal reference for the design was U13, but according to a Product Planning memo of October 23, 1963, the proposed vehicle also was referenced as "G.O.A.T."

Early 1964 prototype of the half-cab Bronco—one of three versions Ford executives planned to launch simultaneously. Note the built-in turntable platform allowing for 360-degree viewing. © 2023 Ford Motor Company.

An early completed 1966 Bronco Wagon displayed (privately) outdoors at Ford's Dearborn Product Development Center. © 2023 Ford Motor Company.

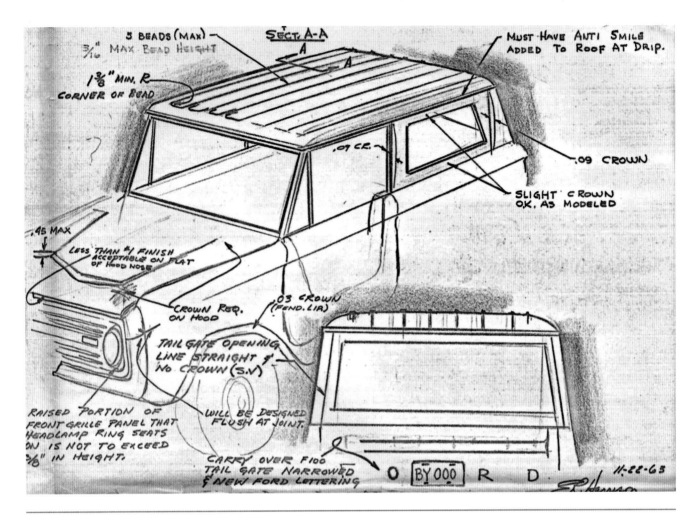

The sketch contains the following handwritten annotations:

- 5 BEADS (MAX)
- 3/16" MAX. BEAD HEIGHT
- SECT. A-A
- A ... A
- MUST HAVE ANTI SMILE ADDED TO ROOF AT DRIP.
- 1 3/8" MIN. R CORNER OF BEAD
- .09 C.R.
- .09 CROWN
- SLIGHT CROWN O.K. AS MODELED
- .45 MAX
- LESS THAN #1 FINISH ACCEPTABLE ON FLAT OF HOOD NOSE.
- CROWN REQ. ON HOOD
- .03 CROWN (FEND. LIP)
- TAIL GATE OPENING LINE STRAIGHT & NO CROWN (S.V)
- RAISED PORTION OF FRONT GRILLE PANEL THAT HEADLAMP RING SEATS ON IS NOT TO EXCEED 3/8" IN HEIGHT.
- WILL BE DESIGNED FLUSH AT JOINT.
- CARRY OVER F100 TAIL GATE NARROWED & NEW FORD LETTERING
- O [BY 000] R D
- 11-22-63

A sketch by E. P. Harrison dated November 22, 1963, provides some insight into the details that were being developed at this point. © 2023 Ford Motor Company.

Not as in "greatest of all time," though that may seem plausible today, but rather "goes over all terrain," indicating the seriousness with which Ford executives were focused on the design's capabilities.

By this point, dynamo Lee Iacocca, Ford's Vice President and General Manager, whose opinions and sage understanding of the automotive consumer were legendary within Ford as well as the industry at large, was toying with both the model names and the intent of this prototype. Iacocca said J. Walter Thompson, Ford's long-standing advertising agency, supplied his marketing team with a series of animal names, which included Mustang, Bronco, Cougar, and Colt. Mustang was adopted as was Cougar. Another story says

stylist John Najjar suggested Mustang based on the World War II fighter plane, the P-51 Mustang. Still another version says that Ford research polled consumers who agreed on Mustang. In any case, Bronco stayed within the family of animals, specifically horses, and the tag earned its place as the G.O.A.T.'s final name.

While marketers worked overtime on the new vehicle's identity, engineers, designers, manufacturing execs, and other key personnel had their own assignments. Many talented design staff were pulling a paycheck from the Blue Oval in the early 1960s, all led by Gene Bordinat Jr., alongside Donald Frey, Philip Thomas Clark, Gale Halderman, and McKinley Thompson Jr., among others.

(continued on page 17)

He was once labeled "Detroit's sharpest idea man," according to *Time* magazine, but Donald N. Frey, PhD, was so much more. The successful third generation engineer was well-rounded in that he was never pigeonholed in one category or career path. Frey started working for Ford in 1951 as a manager in the company's Metallurgic Department, after an education at the University of Michigan, including a doctorate in metallurgical engineering, which is the study and transformation of metals into products for both consumers and manufacturers. He was promoted to director of the engineering research office in March 1957, and six months later appointed Executive Engineer of Ford Division's Car Product Engineering.

His bosses quickly realized the potential Frey had and ensured he remained a loyal employee of the Blue Oval, promoting him multiple times in the next six years, increasing his duties and responsibilities, as the young executive moved from Assistant Chief Engineer of the Ford and Mercury Product Engineering Office to become Product Planning Manager for Ford Division in 1961. Working with Lee Iacocca, who was equally recognized for his contributions, Frey became Assistant General Manager of the Ford Division before he was 40, taking control of all engineering, product planning, and purchasing activities.

It was within this spectrum that Frey, as head engineer, supervised the overall development of the Mustang in 18 months, working mostly in secrecy and prior to official approval, with limited funds. Eventually Henry Ford II, the company's long-term Chairman, green-lighted the project, but known for his bluntness, told Frey he would be fired if the Mustang was unsuccessful. It turned out to be the biggest launch since the 1927 Model A.

Of course, by the time the Mustang was in dealership showrooms, the Bronco was well underway toward its own production date, and Frey had steered its development since the idea first began to percolate around the Dearborn campus. On January 14, 1965, Iacocca, who had served as Vice President and General Manager of Ford Division, was appointed by Henry Ford II to Vice President, Car and Truck Group. By natural progression, Frey succeeded Iacocca as General Manager and was elected a Vice President of the company by the Board of Directors.

In his new role, Frey now earned the right to execute the launch of the Bronco. As head of the Ford Division, he was in charge and oversaw all the details that led to the vehicle's successful launch.

During this time, Frey was also credited with the four-door Ford Thunderbird (fifth generation), the stereo dashboard tape deck, and the popular station wagon multipurpose tailgate that either swung out like a door or dropped down as a conventional tailgate, which first appeared on the 1966 Country Squire and Country Sedan. Ford Motorsports also benefited from his management.

In April 1967 Mr. Ford promoted Frey to Vice President, Product Development, stating in a corporate press release that the position was created to "provide greater integration and coordination of the company's vehicle planning and development activities.

"Consolidation of these activities under a company officer with long experience in the product planning and engineering areas will enable us to apply the most modern planning systems and techniques to a broad program of vehicle development for all of our North American operations," Ford continued.

The appointment had Frey reporting to Iacocca again. The two had caught lightning in the bottle several times over their Ford careers, and as Frey assumed the chairmanship of the Advance Product Planning Committee, there was no doubt more excitement would follow.

Frey moved to head the Product Development Group as Ford further refined responsibilities and departmental functions while the corporation grew and he remained there until resigning from Ford on September 1, 1968, in part due to differences with Iacocca and an opportunity to work outside of the automotive industry. He joined General Cable Corporation, where environmental issues became

Donald N. Frey was a major influence on Ford styling, engineering, and operations. He eventually served as Product Planning Manager before his promotion to Ford Division Vice President and General Manager. © *2023 Ford Motor Company.*

central to his interests. He later was appointed chairman and CEO of Bell & Howell, supporting moves into video cassettes and CD-ROMs in the late 1970s and early 1980s. It was Bell & Howell that developed technology that enabled films to be copied onto videotape in a minute, instantly advancing production, something of great importance to Hollywood.

By 1988, with more than 37 years in engineering and management positions, Frey segued into academia and became a professor at Northwestern University, teaching in the Industrial Engineering and Management Sciences Department. He stayed for 20 years.

When Donald passed away in 2010, one thing everyone who had worked with him could say was that he was clearly brilliant and left his mark everywhere he went and in everything he did. Precision was an appropriate description of his mindset. He was first and foremost an engineer. In 1990 President George H. W. Bush awarded him the National Medal of Technology. But he was always proudest of his accomplishments at Ford, of the Mustang and the Bronco, vehicles that today continue to define the automaker. And for that, Donald N. Frey deserves an incredible amount of recognition and gratitude from both Ford and its customers.

Donald N. Frey (left) and Lee Iacocca. By the time the Mustang launched (technically, as a 1964½ model), Frey served as Chief Designer, while Iacocca was Ford Division's Vice President and General Manager. While there were many people involved in the development and launch of the Mustang and Bronco, both Frey and Iacocca are most closely associated with the two vehicles. © 2023 Ford Motor Company.

McKinley Thompson, shown here inside Ford Motor Company's Advanced Styling Studio where he is working on a sleek prototype, was one of Ford's first African American designers. He contributed significantly to the early sketches of "an open-air 4×4 concept," which became the Bronco. The Art Center College of Design graduate quickly impressed Ford executives, and he remained with Ford until he retired in 1984. He was named to the Automotive Hall of Fame in 2023. © 2023 Ford Motor Company.

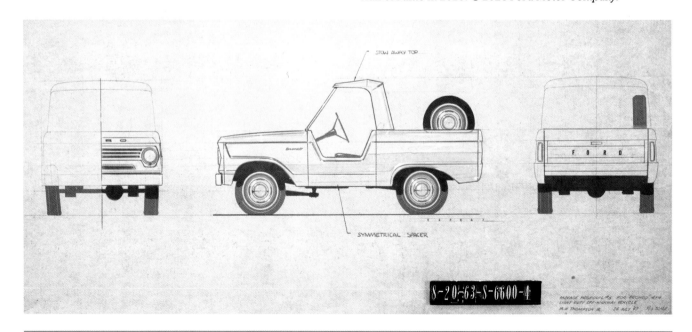

Very early Ford Bronco sketch by McKinley Thompson, July 1963. © 2023 Ford Motor Company.

(continued from page 13)

Meanwhile, Iacocca ran his Fairlane Committee, named after the Dearborn-based hotel where it met, which had been tasked with developing new cars that were younger, sportier, and more fitting for the new decade's expectations. They wasted little time, and the Mustang was launched April 17, 1964, debuting at the New York World's Fair to considerable excitement.

Back at the ranch, as it were, the Bronco had been approved by Iacocca two months earlier. Things were on a roll at Ford.

WHO GOES OVER ALL TERRAIN . . .

How much demand was there for an off-road vehicle in 1964? Who needed four-wheel drive (4WD)? Well, if one lived in any location that endured, or enjoyed, snowfall, then having a vehicle that could overcome this obstacle—and many more—was beneficial.

Farmers and ranchers were embracing new technologies, and a small vehicle that could quickly survey the landscape, scamper after a calf gone astray, or simply (and more comfortably) get from the "back forty" to home was certainly had to have its merits.

But one of the most compelling appeals for 4WD was the growing interest in off-roading as entertainment. Naturally those who had served Uncle Sam and experienced combat duty had seen what a Jeep could do and where it could go. Remote deserts, mountainous sites, and wooded regions were a lure. Fishing, camping, hiking—four-wheel drive made the outdoors accessible like never before.

Thus, the idea of making the Bronco a multipurpose vehicle was critical from the outset. And much like its Mustang stablemate, the Bronco would come in three different versions: a convertible, or more realistically what would be called the Roadster, a doorless, open-top model;

Ford's customers were its dealers, and gauging their interest level was key in generating orders. The Bronco went on a whirlwind tour to build excitement and to highlight the little four-wheeler's unique features. This special intro, complete with cowboy hats, captured the attention of those seated. © 2023 Ford Motor Company.

Donald N. Frey was a major influence on Ford styling, engineering, and operations, eventually serving as Product Planning Manager before his promotion to Ford Division Vice President and General Manager. Here, in 1966, he is posed next to a Bronco engine bay, one of the vehicles with which he was personally involved from its gestation to launch. © 2023 Ford Motor Company.

a two-door pickup (or "Half Cab," as it came to be known); and a three-door wagon, the most common and longest-lasting version.

Models were constructed in 1963, and Donald Frey, who divided his time between the Mustang and the Bronco, shepherded design while Paul G. Axelrad focused on engineering. Many of the earliest drawings were produced by the aforementioned and later by McKinley Thompson Jr., the first African American on Ford's design staff. Thompson was producing sketches of the Bronco by mid-1963. (Thompson was inducted into the Automotive Hall of Fame as part of the Class of 2023. The recognition is "widely considered the single greatest honor an individual can receive in the automotive industry.")

Among the earliest decisions was to build Bronco on a chassis specifically designed for the vehicle. Obviously one requiring four-wheel drive was not available on any other Ford product, despite the manufacturer's penchant for sharing parts within its family of vehicles. The wheelbase was short; just 92 inches (234 cm), with the ORV using a box-section, body-on-frame construction.

Despite the uniqueness of its chassis, Ford accountants were determined to keep costs down, much as they had directed with the early production of the Mustang, though credit had to be given to Frey, who tried to give each vehicle

a complete sense of uniqueness, despite bin-sharing of parts from across the Ford spectrum. Even when the Bronco received widespread approval, conservation of expenses remained the order of the day, and utilizing existing parts off the shelf was not only deemed practical, but necessary.

As part of those cost-saving efforts, as the Bronco moved toward manufacturing, it was deemed that the sole available engine would be the Falcon's 170-cubic inch (cu in) or 2.8-liter (L) inline six. Producing just 105 horsepower (hp) or 77 kilowatt (kW), it was no barnstormer, but it was adequate and, critically, it was reliable. However, this standard one-engine situation lasted only from the initial launch date of August 11, 1965, until March 1966, when an optional 200 hp (147 kW) 289 cu in (4.7L) V-8 became available. Enthusiasts recognized very quickly this engine's source, albeit with minor modifications. The engine wasn't the only component lifted from the Mustang. The bucket seats adorning the first Bronco were taken directly from the stablemate's assembly line as well. Those seats looked right at home in both vehicles.

Most 1960s drivers were accustomed to manual transmissions and, again, to minimize production costs, Bronco only featured a three-speed, column-mounted shifter, along with a floor-mounted transfer case gearshift. Ford wouldn't get around to offering a three-speed automatic until 1973.

Ford's Bronco Roadster, one of three models offered when the Bronco was introduced in 1966, lacked doors and a top, making this a true open-air vehicle. Mimicking its competitors, the windshield even folded down, secured by a bracket on the hood. © 2023 Ford Motor Company.

To keep Bronco's base price low, it was kept as basic as feasible, relying on an à la carte option menu available through Ford or installed at the dealer level. Enthusiasm was high and the vehicle was well-received, with 23,776 Broncos selling in 1966. Mustang, by comparison, moved more than 22,000 cars on its debut in 1964, with a total of 121,538 early '65s, known historically as "64½," between March and August 1964, and incremental sales of 437,913 late '65s through August 1965.

The strategy in launching Bronco mimicked many of the plans and lessons learned in making Mustang a success. Iacocca was the consummate salesman. Drawing on his blossoming public appeal, he worked relentlessly to promote the Bronco, just as he had with the Mustang. Ford's effective public relations team turned up the heat again, and the marketing group and its agency ramped up a glamorous series of print ads for magazines and newspapers nationwide.

Now it was time for customers to respond. To everyone's delight, they did.

In 1954 Toyota adopted the name *Land Cruiser* for its version of a military-type utility vehicle originally requested by the United States Government. Its most popular design and model was the J40 that was built from 1960 until 2001. This is a 1964 Toyota Land Cruiser. *Courtesy of Getty Images.*

Chapter 2

The Overpowering Draw of the SPORT UTILITY VEHICLE

Years have passed since the station wagon was the most valuable tool in the garage. This was particularly true when the family count was on the rise and the wagon carried its weight in groceries, kids, pets, and other trappings, mostly in that order. The popular vehicle that dated back to 1910 handled those chores effortlessly thanks to its row(s) of folding seats and a versatile tailgate conveniently borrowed from its pickup brethren.

For 70 years, the station wagon performed its duties brilliantly, though its popular full-size models with their three rows of seating and reliable V-8 engines began to feel the impact of the first oil crisis in 1973 with yet another salvo in 1979. However, the biggest factor that hastened the wagon's fall from grace was the introduction of an entirely innovative design that completely changed the direction, let alone the purpose, of the vehicle as transportation.

Though the automotive industry has never agreed upon a specific definition, the tag *sport utility vehicle* (SUV) began appearing by the mid- to late-1980s. Adopted by journalists and marketers alike, it became a regular term by the end of the decade. This occurred after a variety of four-wheel drive vehicles began appearing in the showrooms of both American and import manufacturers, the latter including automakers Land Rover, Toyota, and Nissan, all of which had joined the US market either by World War II or in the postwar period.

Toyota's interpretation of the Jeep, referred to as the BJ, later became known as the Land Cruiser, while Land Rover continued to refine its four-wheel drive from spartan off-roader for both consumer and military duty to a more upscale version by the 1980s. The American Motors Corporation (AMC) American Eagle released in 1980 as a wagon with significant ground clearance was certainly a first step toward the SUV as we know it today (though the Eagle

The 1965 Ford Country Squire station wagon with fender-mounted side mirrors and roof rack. Who said man's best friend was his dog? © 2023 Ford Motor Company.

would more likely be considered a crossover in the current market). This was followed by Jeep's Cherokee XJ in 1984, which was actually identified as a sport utility vehicle in marketing materials and corresponding media coverage.

This nomenclature was all decided after the launch of Chrysler's successful minivans, the Plymouth Voyager and Dodge Caravan, introduced in 1983 by then CEO Lee Iacocca, a project he turned into reality after leaving Ford. Iacocca had wanted to introduce the minivan while still at the Blue Oval, but Henry Ford II fired him in 1978 for "personal reasons," ending years of tension between the two men.

But this story is getting ahead of itself. Harken back nearly 20 years, to February 12, 1964, the day Iacocca distributed a confidential Blue Sheet memo to members of the Product Planning Committee, in which he wrote, "Since the introduction of the Scout by International Harvester in 1961 . . . surveys and group interviews [have been conducted] with both Scout and Jeep owners to determine utility vehicle usage and buyer motivation. These surveys indicate that Scout and Jeep owners do not consider their vehicles to be either cars or trucks, rather these units are felt to be especially designed vehicles that can carry nominal loads over all types of terrain."

The memo continued: "The combination of a high degree of maneuverability and a four-wheel drive feature provides an ideal vehicle for use by campers, service station operators, loggers, rangers, and others desiring transportation under adverse circumstances."

Further, Iacocca wrote, "the Bronco will provide a vehicle that combines the best features of both the Scout and Jeep plus requested major improvements in performance, ride, handling, noise, vibration, harshness, and styling."

The executive communication penned by Iacocca was filled with projections, but the most prescient statement was "appearance will become more important as the use of specialized vehicles increases in the expanding recreation market."

At the time, Bronco sales were forecast at 18,000 units annually. The first year saw more than 24,000 vehicles sold.

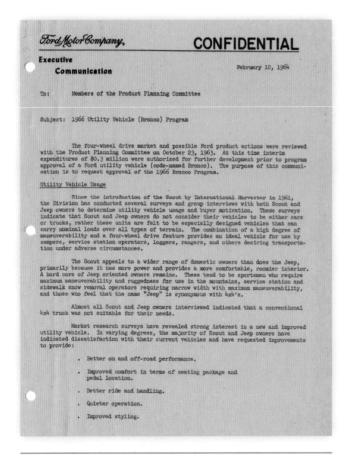

Signed by L. A. Iacocca, the six-page confidential Blue Sheet memo seeking approval of the proposed Bronco project was presented to the Product Planning Committee along with an executive summary, projected sales numbers, product comparison (Jeep and Scout vs. Bronco), financial considerations, and Job 1 expectations. The committee concurred, Iacocca did the final sign-off, and the Bronco was on its way. © 2023 Ford Motor Company.

Always debatable, the AMC Eagle has many historians declaring it the first four-wheel-drive passenger vehicle available to consumers. This is a 1981 Eagle Sport. *Courtesy of Christopher Ziemnowicz.*

Range Rover models, like this 1982 version, were actually imported through the gray market and subsequently modified to meet U.S. regulations. Aston Martin's American operations imported the vehicles in 1984 until Range Rover of North America was formed in late 1985 in order to bring Range Rovers built for the U.S. market through the ports of Baltimore, Maryland, and Long Beach, California. *Courtesy of Getty Images.*

The Jeepster Commando was a quiet offering from the parent company best known for its Jeep CJ, which earned the lion's share of attention and sales. *Courtesy of Getty Images.*

International Harvester's Scout launched in 1961 and served as a benchmark for the Bronco. *Courtesy of Matthias v.d. Elbe.*

International Harvester's Travelall, an early entry in what would become a very large market. *Image courtesy of Mr. Choppers.*

CUSTOMERS CALL THE SHOTS

Bronco was approved within Ford and would live through the 1996 model year, adding millions of dollars to Ford's bottom line and over one million vehicles to customer garages. It eventually ushered in demand for other sport utility vehicles and essentially helped to bury the station wagon, at least as a mainstream vehicle.

There were other participants in the early SUV market, of course. One cannot ignore the International Harvester (IH) Travelall, nor the decades-long existence of the Chevrolet Suburban.

Most vehicles got bigger through the 1970s, with the Bronco taking on girth and width to mimic major competitors, the Chevy Blazer and GMC Jimmy. Jeep was hesitant to change, primarily due to lack of money. Nevertheless, with the Grand Wagoneer, which ran for 28 years, Jeep was one of the first brands to offer a four-by-four as a luxury vehicle.

As the impact of smaller imports grew in the early 1970s, both Ford and GM believed it was a promising idea to launch small wagons. Thus Ford's Pinto wagon and Chevy's Vega (and its sibling, the Pontiac Astre) were released as small two-door station wagons. Meanwhile these vehicles' full-sized siblings grew ever larger, particularly after the federally mandated safety bumpers of 1974. All wagons started looking like tanks ready for battle, but they remained the choice for households needing to haul people and gear. Plenty of baby boomers gleefully recall sitting in the far back seat, facing backward, gazing out the open tailgate window—likely unbuckled—as the wagon rolled along at speed.

Iacocca got the last laugh at Henry II when Chrysler's minivans drove another stake into the heart of the station wagon market. All the manufacturers tried to hijack Chrysler's newfound success, but being first had its appeal and advantage.

(continued on page 29)

1937 Chevrolet Carryall Suburban. Suburban production dates back to 1935. *Alamy Stock Photo.*

How in the world did the sport utility vehicle become the most popular automobile on the road? According to *Autotrader*, the first American-made "crossover" SUV was the 1980 AMC Eagle, the lifted version of the manufacturer's modest wagon applying sibling Jeep technology.

Of course, American servicemembers from World War II will argue that the venerable Jeep they slogged through Europe, North Africa, and the Far East was the first, but purists will likely declare the 1935 Chevrolet Suburban, also was known as the Carryall Suburban, as the original. Interestingly this two-door body style would last through 1967. The first four-door body style was finally introduced in 1973.

Nevertheless most of the original off-road vehicles (as Ford originally referred to the Bronco) were two doors. That remained for decades, though the need for convenient ingress and egress eventually demanded the addition of two doors. Things really began to jump in the late 1980s and early 1990s as indicated in the main story.

But what did come first? Who were the early competitors that convinced the strategists and program developers at Ford that the manufacturer was "leaving money on the table" and needed to launch a multipurpose 4×4 of its own?

A quick glance in the rearview provides some insight and perhaps a bit of hindsight:

International Harvester Travelall

Clearly International Harvester, more commonly known as IH, was a step ahead of its larger automakers when it introduced the Travelall, a big SUV-like model, in 1953. The name was similar to Chevrolet's Carryall, with obvious intentions, as the Suburban was the targeted competition. Offered with either a six- or eight-cylinder engine and built to last, IH pushed hard to make its Travelall a valued choice for big families, businesses, and the buyer who just needed more room. To expand its "shelf space" on the showroom floor, Travelall also featured a sibling called the Wagonmaster that comprised a truck bed and dual cab. Due to demand, Travelall was built for 22 years, ending production in 1975.

International Harvester Scout

Eight years after the Travelall was introduced, a smaller off-roader was launched by IH that started the trend toward more compact and convenient transport. Unveiled in 1961, the Scout became the benchmark, with options ranging from a straight four to a large V-8, a removable hardtop, and like the Jeep, a fold-down windshield. While the Jeep was as basic as basic can be, IH began to focus on luxury and looked to feature elements reflective of that scale early in its production, triggering Ford to take a hard look at the vehicle by 1963.

Though Scout proved popular among multiple audiences, its parent was hurt by a number of factors, particularly a long union strike at the dawn of the 1980s, and Scout's successor was discontinued as a result. The Bronco, however, lived on for another 16 years, though the plucky little truck from IH had made its mark (and will again in the future, this time under Volkswagen ownership).

Jeepster Commando

One cannot ignore the Jeepster Commando, first produced by Kaiser Jeep and subsequently American Motors Corporation (AMC), which had a short run from 1966 to 1973. An intended upscale version of a true off-road vehicle, four models were offered: a pickup, convertible, roadster that featured a removable half- of full-length soft top, and a wagon, with full-length metal hardtop. Though the British glam band T. Rex wrote a song called "Jeepster" (B-side was "Life's a Gas") in 1971, it didn't have anything to do with the attractive Commando, yet young people—aware of the song—flocked to showrooms to have a look at the curious hybrid. Despite strong sales, AMC did not believe the vehicle had a long future and ended its run with the introduction of the larger Cherokee and later, the Scrambler pickup.

Chevrolet Blazer/GMC Jimmy/Dodge Ramcharger/Plymouth Trailduster

Imitation is the sincerest form of flattery, they say. Well, in the automotive industry, if you can't be first, at least be in the race.

It took Chevrolet three years after the Bronco launched to get a similar vehicle onto the market. However, GM's engineers went a slightly different route. They took an existing pickup truck, removed the bed, and installed a full interior with corresponding roof, thus producing a larger version of a 4×4 model. Chevy called its offspring the Blazer while brand-mate GMC received the Jimmy. Customers were delighted to find an alternative vehicle with bigger dimensions and accordingly, bigger engines. Response was good and the two versions remained for sale until 1991. Like the Bronco II, however, GM also sold smaller versions called the S-10 Blazer and GMC S-15 Jimmy. GMC rebadged the full-size Jimmy as the Yukon in 1991 while Chevrolet replaced Blazer with the Tahoe in 1994. The following year, both vehicles earned longer four-door models.

The Blazer—and the Bronco—convinced Dodge and sibling Plymouth that they too had to get into the game. By 1974 Dodge had adopted the same design concept of merging a truck, via a shortened chassis, with a closed full cabin to sell as the Ramcharger and even offered the powerful 440 V-8 as an option for power. Plymouth, as usual, pirated its own version, which it called the Trailduster. It wasn't really all that clever in terms of uniqueness, but at least its dealers got something else to sell in their showrooms.

Ford Centurion Classic, 1987–1996

Naturally the Ford Centurion Classic wasn't a competitor to Ford, but recognized as a licensed, aftermarket production model, what people refer to as a "conversion." The Centurion Classic was created by the Centurion company of White Pigeon, Michigan, whose executives convinced Ford personnel that it could produce and sell an amalgamation of Bronco closed bodies with an F-Series chassis. This vehicle launched in 1987 when bigger became better, and while attractive to many, it was a massive vehicle and took away from the spryness—even when the Bronco became bigger in its later generations—of the original concept. Centurion would pair the full crew cab of an F-Series to the later, larger Bronco body, generating three-row seating. It was the true long-distance cruiser and possibly helped speed up development, or at least consideration, of the Expedition and Excursion officially sold by Ford a decade later.

The licensed Ford Centurion Classic, built by White Pigeon, Michigan's Centurion company from 1987 to 1996, merged an F-150 with a Bronco to build a running prototype of a four-door Bronco, but more likely hastened the imminent Expedition and larger sibling Excursion. *Courtesy of Eric Brandt, AutoTrader.com.*

The Jeep Grand Wagoneer (XJ from American Motors) was initially introduced as a four-door station wagon body style in 1961 and carried a variation of the Wagoneer name to 1991 through three different manufacturers. This vehicle helped create the luxury 4×4 market. *Alamy Stock Photo.*

This 1961 Ford Country Squire station wagon ad illustrated the vehicle's growing size and emphasized its extended warranty to better serve families' expanding households as well as the demands placed on their transportation. *Alamy Stock Photo.*

(continued from page 25)

Ford, however, seemed unfazed by the evolving minivan market. Its F-150 pickup remained on a seemingly endless sales streak (2022 was its 46th consecutive year as the number one truck and it continues as America's best-selling *vehicle* for 41 straight years). The Bronco first gen carried on with a series of upgrades through 1977 followed by an increase in size in its first major restyling for the 1978 model. This version progressed to 1996 with minor changes, primarily cosmetic, though there were engine updates. Sales remained strong throughout the 1980s, only sliding in the 1990s when Ford began producing other vehicles that cannibalized Bronco sales.

Ford's alternate offerings were the result of actions by the company's designers, stylists, and engineers, who sensed a shift building with key consumers (think moms, business-people, and taxi drivers, among others), all of whom had been driving a minivans for a decade and were tiring of both the vehicle's format and its image. The time seemed ripe for something new and not unlike the vehicle the Bronco had originally established: "vehicles that can carry nominal loads over all types of terrain." In other words, a vehicle for stuff, people, pets, and products—but different.

The Ford Explorer arrived on dealership lots at a time when comfort, size, and durability ruled America's roadways. The model helped popularize the sport utility vehicle category and appealed to those who wanted to pursue their own paths and carry plenty of passengers and cargo. © 2023 Ford Motor Company.

THE POWER OF PROMOTION

What else contributed to an increase in demand for "alternative" vehicles?

Advertising definitely played a role. The launch of the Jeep Cherokee (SJ/XJ) in 1974 (owing its heritage to the Wagoneer, and later Grand Wagoneer) helped usher in both a new era and a new segment: a two-door or four-door vehicle offering practicality (especially for winter climates) and off-roading ability. The gauntlet was thrown to other manufacturers.

As mentioned earlier, Land Rover began to up its game, pursuing higher income buyers with more elegant offerings. They had only one upscale challenger for a brief period, a hybrid Italian-American former troop carrier called Laforza. Designed by renowned stylist Tom Tjaarda (of DeTomaso Pantera fame) for the Italian Army, the stylish 4×4 came to America in 1989 cobbled together from a variety of parts, including the Bronco engine! The vehicles were assembled by Cars & Concepts (later called C&C) at their facility in Bristol, Michigan. Its dealer network was an amalgamation of Italian sports car reps, an occasional Ferrari store, and independent shops. Priced at $50,000 in 1990, it was an expensive, but similarly priced, counteroffensive to the Range Rover. It should have had a better shot, but succumbed due to a lack of money and sales outlets.

By 1990 Ford was readying the Explorer, the carmaker's first wagon-style SUV, which, under the direction of Ford's Truck Package Team, led by Ben Callaway and Jim Rowlands—according to retired Ford Chief Designer Bob Aikins—proved to be exactly what consumers were looking for at that moment. The 1991 model, the first of its kind, launched in March 1990, sold 140,509 units with another staggering 282,837 models throughout 1991. Intended to replace the Bronco II, which offered only two doors, the new Explorer offered both two- and four-door configurations (technically, three- and five-door versions when considering the full rear liftgate with a flip-up window). That first-generation Explorer sold more than 300,000 vehicles per year—outselling the Jeep Grand Cherokee and Chevrolet S-10 Blazer by a sizable margin—and more than 400,000 units per year by the end of its second generation, becoming the ninth best-selling vehicle in the United States by 1994.

The Bronco II was an attempt by Ford to offer a second SUV with the Bronco name that was similar in size to the first-gen model. Despite Ford's best intentions, the Bronco II, based on the Ranger compact pickup, lasted only from 1984 to 1990 and was replaced with the Explorer upon its launch in 1991. © 2023 Ford Motor Company.

Retired Ford Chief Designer Bob Aikins was charged with creating the Explorer, positioning it to replace the Bronco II, but originally as a four-door only. However, prior to final production sign-off, Aikins was asked by Ford executives to add a two-door model. This 1990 two-door Explorer prototype was production ready. © 2023 Ford Motor Company.

Interestingly the first Explorer shared much of its infrastructure with the Ranger pickup, as well as the Bronco II. Ford's marketing touted the vehicle as being "family-oriented" while possessing off-road capability, thus increasing its appeal across incremental markets. That, and the four-door option, neatly attracted buyers. Prior to becoming Ford's Chief Designer in 1989, Aikins's first project as the Design Manager in the Truck Studio in 1984 was the new Explorer.

"The Truck Studio was very small in those days; I was the exterior manager and there was one interior manager," he said. "We had a Director, John Aiken, but no Chief Designer.

"The Explorer was indeed positioned to replace the Bronco II," Aikins continued. "I remember a widened rear track was a big deal and the vehicle was originally designed as a four door only. I initially had a very small exterior design team with Ken Saylor and Neheniah Amaker. The design was not a clean sheet project as we were forced to use a lot of Bronco II parts to keep the initial investment low. Nobody at Ford knew or expected the huge sales success, something, in my opinion, that was due to Callaway and Rowlands, in part because there was no real four-door SUV on sale at that time, including the four-door Jeep, which we felt wasn't competitive, plus the great package these two guys created."

Aikins went on to add that at the time his team was dealing with Explorer, "we also had the design responsibility for the F-Series and big Bronco, plus the extended Aerostar and the Aeromax heavy truck; we were busy bees!"

Pleased with the results of the finished four-door Explorer, Ford Division came to Aikins and requested a two-door version prior to production sign-off. According to Aikins, "I took a four-door Explorer fiberglass mock-up, and with Jim Rowlands, dimension-sectioned the vehicle and magic! It became a two-door."

Obviously, that was a brilliant move on management's part as the two-door clearly retained a market, and buyers who had been used to a two-door smaller off-road model via Bronco II had a new, updated choice.

"In my view," Aikins resumed, "the impact of the Explorer on sales of the big Bronco had more to do with buyer preference. Our mid-sized SUVs were replacing the aging 'soccer mom' minivans."

Now well into the vehicle's sixth gen, which commenced in 2020, the 2022 Ford Explorer ST is the marque's performance-oriented model. The entire Explorer line has been a tremendous success for the Blue Oval, with sales now totaling more than 8,400,000 units since its 1990 launch. © 2023 Ford Motor Company.

BRAGGING RIGHTS AND BRAND EXTENSIONS

Six generations and 30 years later, the Explorer continues to bring home Ford's bacon, having thus far sold roughly 8,300,000 consumer models and another 151,212 police interceptors. Spin-offs have included the Mazda Navajo, Explorer Sport Trac, Mercury Mountaineer, and Lincoln Aviator, plus significant export sales.

While Explorer sales grew, big Bronco sales began to decline, though its final years were significantly higher than 1992, the first year of the fifth-generation design. Nearly 25,500 big Broncos sold in 1992, and annual units ramped up to more than 34,000 by 1996, its last year. That said, 402,663 Explorers sold in that same year. The writing was on the wall for Bronco.

The Expedition debuted in 1997, a larger four-door SUV and intended, though not necessarily publicized, as Bronco's successor. It was Ford's first full-size sport utility vehicle and based on the corresponding generation of the F-150, shared both body and mechanical components.

As Chief Designer, Aikins had overall styling approval for the Expedition, which was completed in early 1993, three-and-a-half years ahead of its planned 1996 production date. Chief Engineer Dale Claudepierre had responsibility for the SUV's development, and the entire budget for development was set at $1.3 billion.

Announced on May 9, 1996, and launched October 2, 1996, as a 1997 model, the Expedition offered optional three-row seating, along with a multitude of options. During the calendar year 1996, Expedition sales numbered 45,974, climbing to 214,524 through 1997. Ford had yet another hit on its hands.

Lincoln got its own version, the Navigator, in 1994, overseen by designer Fritz Mayhew, which launched on July 1, 1997, one year after its Expedition cousin. Ironically Jeep had stepped away from the luxury market

The 1997 Expedition Eddie Bauer model, its first year of production. Ford designers and engineers packed this big SUV with so many goodies, including optional three-row seating, that it quickly became the rig of choice for big families and replaced what the Bronco might have become. © 2023 Ford Motor Company.

On the heels of the Expedition came the Excursion because, well, bigger is better. The first model debuted in 2000, the longest and heaviest SUV to enter mass production and aimed squarely at the competition, mainly the ones wearing a bowtie. *Courtesy of Getty Images.*

with the demise of its Grand Cherokee in 1991, leaving the door wide open for a higher-end brand to step in and grab market share. Oldsmobile had tried with its 1990 launch of the Bravada, but didn't create much noise. Thus Navigator jumped right in and for the first time in decades, Lincoln overtook Cadillac (which wouldn't release its Escalade until 1999) in annual sales volume, a considerable bragging right.

As more SUVs entered the marketplace, particularly among the imports, going big became the norm. Right before the world transitioned to a new century, Ford happily ushered in the Excursion on September 30, 1999, as a 2000 model. It was the longest and heaviest SUV to enter mass production and a direct competitor to the larger Chevrolet Suburban and GMC Yukon XL. Though Excursion had a short lifespan—suffering criticism for its size and low fuel economy—sales still met expectations.

During its seven years of production, Excursion sales totaled roughly 200,000 vehicles. It was retired in 2005, but was quickly (and subtly) replaced by the new third generation Expedition in 2007 through a long wheelbase version called Expedition EL in the United States and Expedition Max in other markets. In a smart move, the Expedition was classified as an ultralow emission vehicle (ULEV), blunting the previous reproaches for being a gas guzzler. As a result, Expedition continues to successfully compete with the larger Suburbans and Yukons.

Ford maintained a strong percentage of the SUV market share and showed additional promise in a newer line of crossover utility vehicles (CUVs) by introducing the Edge in 2007. This was a further sign that the sedan's days were numbered, as the Edge shared its platform with the Fusion and formed a neat position between the Escape (launched in 2000) and Explorer in Ford's product plan.

Priced right and packaged well, Edge has served its purpose as a durable, middle-of-the-road family vehicle. It has spawned numerous international versions as well as a cousin for the Lincoln brand, first known as the MKX and more recently as the Nautilus. Though earmarked for its final run in 2024, Edge has been a strong sales performer, notching over 100,000 units every year except 2009 and 2021, totaling close to two million domestic sales, plus another half-million models in China, Europe, and Mexico.

Despite the strength of this new lineup, Ford desired to retain a vehicle that remained within the wagon category, while also offering the improved fuel economy and ride quality characteristics of a crossover. One classic example was the Flex, produced for one generation from 2009 to 2019. It was a replacement for the Ford Taurus X and Freestar Minivan. A long five-door wagon, its unique design featured a "floating roof," usually in white with all pillars painted black, plus horizontal grooves in the doors and tailgate. Flex was met with approval by a specific element of the car-driving audience who wanted something different and without the height of the Explorer or other SUVs. Over its lifespan, Flex sold more than 307,000 vehicles.

Finally, on an even smaller and more economical scale, Escape debuted in 2000 as a compact SUV, though some referenced it as a crossover. Twenty-three years later, Escape is now in its fourth generation. It also is sold as the popular Kuga outside North America and has generated 4,420,173 sales domestically with another 50,000 buyers in Mexico and Australia, while sister Mercury Mariner captured another 200,000. European sales, as the Ford Kuga, have tallied another 1.4 million, while Chinese purchases of the Kuga added up to 630,000 (the vehicle is now sold as an Escape in China). Over its life, the vehicle was also branded as Mazda Tribute as well as Maverick in Europe and China. It clearly has been classified as another successful model in Ford's SUV lineup.

While people's taste in transportation has changed over the last few decades, and North America in particular has seen a vast transformation within the automotive landscape, the fact remains that personal vehicles typically serve multiple purposes, and one of those purposes is the conveyance of more stuff than just people. The station wagon was a loyal servant for more than 70 years, but like many products that become obsolete thanks to technology or trends, it fell by the wayside, superseded by minivans, pickup trucks, and ultimately, SUVs. Though most drivers rarely take their four-by-four off road, they value the comfort and security of having all-wheel drive. Others simply prefer the safety and control of sitting higher. The market victim is anything "car-like."

The future looks bleak for "normal" cars from the Blue Oval (and other manufacturers). It's trucks, SUVs, and CUVs, save the hearty, durable legacy that is Mustang. Will that work? Why, yes . . . yes it will.

Priced right and packaged well, this 2022 Ford Edge has been very successful in its role as a durable, mainstream family vehicle. The Edge was Ford's first mid-sized crossover utility vehicle (CUV). It has spawned numerous Ford Motor Company domestic and international versions with a third gen recently launched for China. The U.S. model will end in 2024. © 2023 Ford Motor Company.

Going to the opposite extreme of the Excursion, Ford served up its smallest SUV, the Escape, in 2000. The versatile economy crossover has created a true niche within the Blue Oval line-up of vehicles, closing in on five million sales in North America and worldwide. © 2023 Ford Motor Company.

Ford designers had tweaked all the lines, features, and
components by the time this Bronco prototype appeared
in early 1966, save for wheel caps. © 2023 Ford Motor

Chapter 3

The First Bronco, 1966-1977

On Halloween day, 1963, a Ford intracompany communication, still referring to what would become the Bronco as the "1966 G.O.A.T.," discussed a feasibility clay model review held at the Styling Studio, with the promise that a follow-up would occur one week later in the Rotunda to examine revisions.

Thus followed a number of meetings, viewings, discussions, revisions, photography, feasibility studies, budget requests, allocations and confirmations, and many late nights. Were there arguments? Does Ford have a Glass House? One thing was clear—passion overruled convention. Ford would get its G.O.A.T.

The renowned confidential Blue Sheet memo was issued on February 12, 1964, urging the Product Planning Committee to approve the 1966 Bronco Program. At the bottom of the six-page document was the straightforward summary: "The proposed 1966 utility vehicle program will provide Ford with a superior entry in the utility vehicle market and an opportunity to increase market penetration and Company profits through incremental sales.

"The concurrence of the Product Planning Committee in this program is requested."

It was signed by Lee Iacocca.

Once agreed, the Bronco moved forward, and the $300,000 that had been expended thus far in development costs was significantly raised—as was the bar—to produce the kind of sales winner the Ford Division Vice President promised. Iacocca delivered, just as he had with Mustang, adding to Ford's coffers, launching Bronco's legacy, while adding to his own.

PUBLIC AWARENESS

A Ford press release issued prior to production (on April 28, 1965) spelled out basic Bronco information:

"With its standard two- or four-wheel drive, the new Bronco is equally at home on rugged mountain grades or on a run to the shopping center." (During an interview with Ford Archives Director Ted Ryan, he made it clear that even in that era, Ford marketing personnel were smart not to target a gender.)

Marketing personnel got busy preparing materials to promote the new Bronco featuring the Roadster and Wagon and the latter's towing capacity in a trio of photos demonstrating the various configurations the new SUV could assume. © 2023 Ford Motor Company.

The press release continued: "The roadster is the 'basic' Bronco. It has neither cab roof nor doors and is an open sports model with a windshield that can be folded flat and secured to the hood. A bench seat is standard.

"Adding full doors with roll-up windows and a bolt-on steel driver's cab converts the roadster into a weather-snug and lockable sports-utility model. A full-length steel roof, bolted to the body sides and windshield, turns the Bronco into a fully enclosed station wagon.

"Four-passenger seating is achieved with a two-passenger bench-type rear seat in combination with front bucket seats. In this arrangement the steel bulkhead is deleted to permit access to the rear seat."

The release goes on to discuss the 170 cu in (2.8L), 105 hp (77 kW) six-cylinder engine "specially adapted for rugged, off-highway operation," coupled with the manual three-speed, "synchronized transmission [which] permits shifting into low gear without stopping on steep grades."

Ford took immense pride in its transmission design and engineering: "Shifting into or out of two-wheel or four-wheel drive is managed easily by use of a floor-mounted shift lever. It is not necessary to stop or declutch when shifting into or out of four-wheel drive in high gear. The Bronco's transmission also is designed for an optional 'power-takeoff' to power equipment such as logging saws."

And finally, the release concluded with "two axles are offered on the Bronco—standard and heavy-duty. Both are available with limited-slip differential for maximum traction. Additionally, the front suspension combines extreme ruggedness with excellent anti-dive characteristics—even under panic-stop conditions."

Within that earlier release, Don Frey was quoted as saying that Bronco would draw on the production experience Ford had in designing and building the M-51—a military utility vehicle more commonly known as the MUTT—during the late 1950s and early 1960s, along with more than 282,000 four-wheel drive military vehicles built by Ford during World War II.

Over time, test drives, and design changes, the little vehicle began to take on a life of its own. At the Bronco National News Conference (August 10-11, 1965, at the Kingsley Inn in Romeo, Michigan), journalists got their first thorough look at the new model.

By this point, Donald N. Frey was General Manager of Ford Division and a Company Vice President. He was the featured speaker on August 11th. He announced that Ford Motor Company's Bronco, a completely new line of sport utility vehicles for 1966, would join the popular Ford Mustang in offering active people new driving adventure with one significant difference: the Bronco operated just as effectively off the highway as on it.

Bronco was neither a conventional car nor truck, Frey stressed. It was a vehicle "which combines the best of both worlds.

"Designed as a go-nearly-everywhere, do-nearly-everything vehicle, the versatile Bronco comes in three body styles—open roadster, short-roof sport utility, and fully enclosed delivery or station wagon," Frey explained. "Each model is designed to combine outstanding performance and comfort with practicality."

The goal was to offer thrice as many vehicles and triple the sales opportunities. With multiple products, Bronco had to appeal to a wider audience and succeed in bringing new customers into Ford dealerships.

Frey noted there were 300 "organized four-wheeler groups in the U.S., numbering about 10,000 members, who use utility vehicles for driving and camping in rugged locales.

"We talked to these owners," he stated, "who enjoy exploring the remote wilderness and they told us what features they wanted most in a sport and utility vehicle."

To further support Ford's investment in Bronco, Frey cited that the market for utility vehicles had grown from slightly more than 11,000 in 1960 to more than 40,000 in 1964, and he forecast demand would expand to 70,000 by 1970. "Bronco will be a leader in that growth," he predicted.

"In the past five years, we at Ford Division have introduced three new vehicle lines—Falcon, Fairlane and Mustang. Today, we add a fourth, Bronco.

"We have designed and produced each of these new lines to serve the specialized wants and needs of our customers," Frey went on. "And so far, our efforts to tailor our products to customer requirements have been outstandingly successful. The Bronco represents our latest efforts to tailor a vehicle to the specific requirements of a large and growing segment of the motoring public. Like its older brother, the Mustang, it will be offered with a wide range of options and accessories that will permit it to be many things to many people."

In closing his time at the press conference, Frey stated, "We believe the Bronco will offer customers new standards in this type of vehicle including ruggedness, maneuverability and 'go anywhere' roadability."

An accompanying document highlighted Bronco's features and then laid out options available for customers that could be dealer-installed. These included some pretty interesting choices:

Upgrades are the name of the game, and Bronco came with plenty of them, even in its first model year. Every dealer loves the opportunity for an upsell. © 2023 Ford Motor Company.

- Air springs—overload front
- Compass
- Convertible top—vinyl full-length
- Doors—vinyl with plastic windows
- Fire extinguisher
- Locking gas cap
- Chrome handrails
- Front hubs—Warn Industries free-running
- PTO (Power Take-Off)—front and rear drive
- Snow plows
- Tachometer
- Tool kit
- Tow hooks—two designed for the front
- Trailer hitch
- Two-way radio
- Front-mounted winch

As the press conference continued, Ford Division's General Marketing Manager Walter T. Murphy took over the podium for a spirited question-and-answer session, highlighted by his declaration of Bronco's greater ground clearance, something 4×4 owners expressed they wanted, followed by a rally that began in the Kingsley Inn parking lot and ended with a "Bronco Busting Program" at the Romeo Proving Ground. Journalists who participated in the press junket indeed got the hands-on experience with the new vehicle they were craving and much of their subsequent reporting in the car enthusiast books and newspaper articles was positive. Bronco had arrived.

CONSUMER REACTIONS

Early advertising reflected its versatility. The message was unequivocal. It stressed that Bronco was adaptable to one's personal lifestyle. True, Bronco was comfortable about town, on the trail, and with some upgrades, winning off-road races, like Baja.

By March 2, 1966, Bronco received the 289 V-8 as that appropriate upgrade, making the vehicle's horsepower the class leader among its competition. It was a brilliant move that attracted buyers who craved the additional muscle Bronco now possessed.

At $2,369 for the base model, Bronco exuded simplicity and economy. It was cost-efficient to produce with parts pulled from other factory bins: brakes and axles from F-100 pickups, with the front axle mounted using radius arms

Take a short wheelbase, a lightweight two-door ORV (off-road vehicle), and stuff the engine bay with Ford's bulletproof 298 V-8 and what does one have? Instant success, that's what. A powerful, hill-climbing, cattle-wrangling, highway-embracing all-around sports wagon. © 2023 Ford Motor Company.

and a lateral tracking bar to center the housing under the frame. This setup allowed the Blue Oval to use coil springs that gave the Bronco an impressive 34-foot (10 m) turning radius. A Dana 30 differential was installed first and used through 1971 when an upgrade to the Dana 44 was implemented. Three years after the 289 was introduced, a 302 Windsor V-8 was installed, garnering noticeable respect.

Motorsports has a way of leveling all playing fields and in proving the superiority of design, engineering, technology, and competitive spirit, and Ford confidently displayed each. As Ford demonstrated that four-wheel drive was one of Bronco's unique selling propositions, the sport utility vehicle was immediately popular among those favoring trail-running and off-road racing.

Shift to Long Beach, California, and Bill Stroppe began converting Broncos into off-road racers, which quickly began winning, first at the 1967 Riverside Four-Wheel Drive Grand Prix, followed by victories in 1968 at Riverside, the Mint 400 in Las Vegas, and the highly respected Baja 1000 (originally called the Mexican 1000) where Larry Minor and Jack Bayer drove a stock Bronco to victory in its class. The very next year, Minor and Rod Hall took the class and overall victory as their Bronco was the first vehicle of any kind to cross the finish line.

Quickly, the Stroppe-prepared Broncos were generating national attention and appealing to racers of all caliber and skill, including actor James Garner, who had earned his racing props from the popular film *Grand Prix* directed by the renowned John Frankenheimer, and attracting the attention of Indianapolis 500 winner Parnelli Jones.

Jones partnered with Stroppe and made his off-road bones through the Stroppe-modified 1970 *Big Oly*—thanks to sponsorship from Olympia Beer—and by winning the 1971 and 1972 renamed Baja 1000 (due to a change in sanctioning bodies) races. Their victories became Bronco successes and helped justify the vehicle's four-wheel drive capabilities. Racing has a way of doing that when the brand is winning.

It's October 1966, and a resident of Southern California is walking down the strand in Redondo Beach. He spots this short wheelbase, open-topped sports vehicle parked in front of a popular burger joint, and thinks this smart, beautiful red spark plug just looks like it's ready to tear up U.S. 1 toward Malibu. With so much space available in the back, the local realizes a surfboard could fit just fine. The name on the fender says "Bronco." The seats are silver vinyl and can be easily wiped clean to get the sand out. The license plate bracket states Cal Worthington Ford. He's the dealer with the "My Dog Spot" TV commercials. Everyone has seen them. Since Long Beach is pretty close by, it might be worth it to go see how much this Bronco costs. It would be a pretty cool utility vehicle to own. All the surfers will want one.

The Stroppe-built Bronco named *Super Pony* took the overall win at the 1969 NORRA Baja 500 with Larry Minor and Jeff Wilson on board. *Photo by Pat Brollier/The Enthusiast Network/Getty Images.*

1963 Indy 500 winner Parnelli Jones enjoyed a second racing career as a champion off-road driver, partnering with car builder Bill Stroppe and the Ford Bronco. Parnelli won the NORRA Baja 1000 twice and also claimed victories in the NORRA Baja 500 and Mint 400 in 1973. *Photo by ISC Archives & Research Center via Getty Images.*

The legendary *Big Oly*, so-named due to its Olympia Beer sponsorship but also because it was an imposing and incredibly fast Bronco, especially at the hands of Parnelli Jones, gets a quick diagnosis by the team prior to running the 1972 NORRA Mexican 1000. *Photo by Eric Rickman/The Enthusiast Network/Getty Images.*

The impression this Roadster must have made . . . or for that matter, that *all* the versions of the new Bronco must have imparted. It was such a bold move. It's possible that it wasn't the safest vehicle on the planet, but it was attractive, cool for its time, and about as versatile as a vaudeville act. And like the Mustang, Ford had three models to offer the demanding, yet discriminating, buyer.

The original Bronco was built on a box-section chassis with a 92-inch (234 cm) wheelbase, a live-axle up front with coil springs, and a live axle at the rear with leaf springs. The first engine was a modest six tied to a three-speed manual, fully synchronized transmission, and two-speed transfer case delivering 4×4 capability part-time. The PTO, or power take-off option, enabled dealer-installed accessories such as front-mounted winches, tow hooks, and snowplows to be used when appropriate.

While the launch of Bronco, noted *Automobile* Writer David Zenlea, wasn't the first vehicle of its kind, "it was the first to hint that the segment—which in the mid-1960s was quite small—had a lot of potential."[1]

The Ford Bronco was brilliant and simple: roofless, doorless, lightweight, and with a windshield that dropped down. This is about as close to unfettered driving as one can get. *Courtesy of Winning Makes, Santa Barbara, California.*

Ford recognized that potential early on. The confidential Blue Sheet memo from Iacocca predicted bigger sales that proved to be conservative at the time as Bronco gradually beat those numbers, selling more than 94,000 versions by 1970, its first half-decade of production.

Who bought them? Ford called Bronco a "new kind of sports car with four-wheel drive." The open wheeler attracted those who cherished sunshine. Four-wheel drive brought in cold weather cowboys, mountain climbers, and recreational enthusiasts. Stroppe drew the growing clan of racers. And the overall appeal of the "go over any terrain" and "fit any lifestyle" increased the desire among farmers, ranchers, campers, and those who wanted a different style of "family station wagon." Frey called them "about as solid and respectable a segment of the automobile market as you will find anywhere in the country."

Of course, this didn't include government purchases for state parks, federal highways, police, and civil defense work. The Bronco logo was appropriate as the little vehicle became quite a workhorse. Even the military wanted a share.

As Walter T. Murphy, the head of marketing, confirmed during the media conference and vehicle introduction, Bronco offered "a unique combination of ideal wheelbase, road stance, tight turning diameter, smooth ride, ruggedness, capacity, comfort, convenience, power, and utility."

Initial weight was 2,780 pounds (1,261 kg), increasing over Bronco's first-gen life. Plus the addition of safety, smog, and other equipment, as well as the availability of a 289 cu in (4.7L) V-8, putting out 150 hp (110 kW), and later a 302 cu in (4.9L), capable of 158 hp (116 kW), plus a three-speed automatic.

Interesting highlights from Ford's announcement during its launch included mentioning Bronco's anti-dive front suspension as well as an optional 11-gallon (42L) gas tank mounted under the driver's seat with a switch below the dash controlling the fuel gauge readings. To top things off, Bronco came with a 24-month or 24,000-mile (38,624 km) warranty (whichever came first, of course), which was the best warranty of its kind on the utility vehicle market.

Frey declared in his closing remarks that "to fill initial demand, we plan to build 18,000 units by the end of December [1965] at our Michigan truck plant. We think our dealers will need every one of them."

Truer words were never spoken.

Bronco was getting all kinds of looks and identities as the vehicle began to build in popularity. As a Roadster, Ford even went so far as to call it a "new kind of sports car." And why not? Open top, no doors, big motor? And off-road capability too? Let's go! © 2023 Ford Motor Company.

1967

Fast forward to the 1967 model year and Ford advertising is positioning Bronco as the manufacturer's "gift to the year-round sportsman."

"You name the place and Bronco will take you there . . . a get-away-from-it-all oasis off the beaten paths."

The versatile four-wheel utility is engineered for such trips, Ford maintained, and yet, will take its owners over "today's superhighways in passenger car comfort."

This could have been the Explorer Ford marketers were talking about, but this was 23 years earlier.

Bronco was a good thing and so amendments were scarce. Ford did decide to introduce two new models, the upgraded Sport Bronco Wagon and Sport Bronco Pickup,

which featured interior and exterior trim appointments including bright metal grilles, bumpers, and bumper guards; bright, full wheel covers; luxury vinyl door trim; and bright metal interior accents. The most distinctive feature of the Sport model was the red-painted *FORD* grille lettering.

As with most of Ford's vehicle lineup, reverse, or back-up, lights became standard this year.

From a performance perspective, Bronco offered an optional 11.5-gallon (43.5L) auxiliary fuel tank, mud- and snow-type tires for 15-inch (38 cm) wheels, a dual brake system with warning light, improved seat anchorage and door latches, and a new floor-mounted T-bar shift lever to operate the transfer case. Both the 170 cu in (2.8L) six and the 200 hp (147 kW) 289 V-8 were offered. Ford also was enormously proud of its specially designed two-venturi carburetor for steep-grade performance, special six-quart (5.7L) oil pan

for hill-country lubrication, no-spill oil-bath air cleaner, full-flow oil filter, and fully aluminized muffler and tailpipe.

1968
Perhaps the most dramatic upgrade and most distinctive change in profile for the 1968 Bronco was the optional swing-away spare tire carrier that was hinged to a mount on the right rear corner of the body. A spring-loaded slam latch secured the spare to the tailgate. The decision among stylists and marketers at the Blue Oval was to give Bronco more convenience for accessing the spare as well as much-needed extra cargo space. The exterior-mounted spare also largely contributed to the overall look of Bronco and ushered in a second wave of customization and incremental sales for Ford, its dealership body, and the burgeoning aftermarket—the spare tire cover.

Ford decided to add several safety-oriented features to the standard Bronco interior. This is September 1967, just two years after Ralph Nader's critical examination of the automotive industry *Unsafe at Any Speed*, in which he accused auto manufacturers of their reluctance to invest on safety improvements to their vehicles. Hearings conducted by the U.S. Senate engendered by Nader's book led to the creation of the U.S. Department of Transportation in 1966.

Those interior upgrades included standard armrests on all models with doors, produced as a new yielding design. Door handles featured a new recessed flipper grip for greater safety and easier operation while front lap seatbelts became standard.

Thus, safety advances and additions became more commonplace among all Ford model lines, as well as those within other original equipment manufacturers. On Bronco, new free-running front hubs with improved lubricant sealing and simpler operation were installed as well as a new kingpin with a high-density polyurethane-filled bearing cap that automatically compensates for kingpin wear and provides improved anti-shimmy capability and virtually service-free operation for the life of the vehicle.

In 1968 the now year-old Bronco Sports had vinyl-covered door trim panels with bright moldings, fitted vinyl floor mats, horn rings, cigarette lighters, and many other interior and exterior luxury touches. Bumpers also had curved ends to them along with new side marker reflectors.

This year featured a number of safety improvements. For example, 1968 meant the introduction of the "Ford Motor Company Lifeguard Design Bronco Safety Features" program. The Roadster, as with all vehicle production, had to follow the evolution of guidelines implemented by the U.S. Department of Transportation, which likely ushered its imminent demise. The dual hydraulic brake system with a warning light and 4-way emergency flashers became standard. For all Broncos, a new and improved energy-absorbing pad, safety-designed control knobs, and a taller rectangular speaker grill were all improvements made in the dash, which was changed to steel.

One mandatory change to the Roadster body was the addition of side body reflectors. Grille and taillight reflectors could be installed as factory upgrades as well. Gone were the solid red taillight lenses of the '66 and '67 Roadsters as integrated backup lights became standard production. The grille's clear turn signal lenses vanished after 1968, replaced with oversized orange lenses. Visibility was key.

Finally the two engine choices were now fitted with dry air cleaners, eliminating the oil bath air cleaners.

The 1968 Bronco Roadster remained relatively unchanged from its launch, a beautiful open-air, door-free runabout that exemplified the moment—freedom of movement and expression. Unfortunately due to dwindling sales this was Roadster's last year. © 2023 Ford Motor Company.

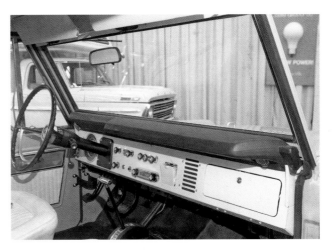

The North American International Auto Show, a.k.a. Detroit Auto Show, has always been one of the grandest new vehicle shows in the world, and Cobo Hall was awash in chrome and steel every January for decades. The 1968 Bronco was a star as this interior shot shows within the Ford booth. © 2023 Ford Motor Company.

1969

After minimal sales over its three years of production, the last a disappointing 212 units, the Bronco Roadster was discontinued. Officially 5,000 of these doorless, topless runabouts were sold during its short existence.

The second most notable change was the 302 cu in (5.0L) V-8 replaced the venerable 289 and Bronco's fenders proudly displayed those new engine emblems.

In addition, electric wipers became standard and visible to consumers, the instrument cluster design was revised. What was not visible but perhaps noticeable to owners was that Dana's Spicer brand axle and driveshaft products became the original equipment (OE) for Bronco, replacing the Borg Warner driveshaft.

Customers who ordered the Sports model received aluminum door panel trim, a pleated parchment interior, and a rear floor mat when the rear seat was optioned. Amber-colored parking lights lenses replaced the original white ones that were harder to see.

In 1969 Bronco received a nice engine upgrade when its powerplant moved to the 302 V-8, and the front fenders earned badging declaring such. © *2023 Ford Motor Company.*

1970

The dawn of a new decade brought relatively minor revisions, and most were safety oriented. Of the most visible, taillight reflectors changed along with repositioned side marker lights.

It's tough to change a winning combination, but in the auto business, every new model year requires some upgrade and new modification to bring in both new buyers and existing customers who want the latest and greatest. By 1970 Bronco was getting more sophisticated with increased safety equipment. Building upon the successes of its predecessors, Bronco was on its way to becoming a solid SUV in the marketplace. © *2023 Ford Motor Company.*

1971

Ford strengthened its relationship with Dana as it did with Bronco, swapping out the supplier's original Dana 30 front axle for Dana's stronger 44 component, giving the vehicle more confidence off road. The Dana 44 remains one of the toughest axles available and is a strong choice for rock crawlers and even drag racers because of its durability in handling both power and strain. Clearly Ford engineers understood their audience, as they also moved the fuse box into the glove compartment with a new wiring harness that replaced the old three-piece unit. A heavy-duty radiator also became an option.

Other upgrades that could be ordered separately included a remote control left-hand outside mirror as well as a new headliner for the pickup only. Interestingly, the 302 emblem was removed.

This was the first year for the limited edition (a.k.a. special edition) Baja Bronco built by Bill Stroppe and Associates, which could be ordered through Ford dealerships (see Chapter 13 for more details).

Off-road racer/builder Bill Stroppe quickly adapted into custom Bronco builder, too, and turned his Long Beach, California, shop into the center of off-road modification with the street-legal Baja Bronco, which he offered to enthusiasts and racers alike. Its orange, white, and blue color scheme was just part of the exciting package Stroppe offered. © 2023 Ford Motor Company.

1972

Sadly after selling fewer than 2,000 vehicles annually for the last two years, the Bronco Pickup, also known as the Half Cab, was discontinued, leaving only the Wagon. However, Ford was content for the moment with keeping just the Wagon along with its Sports offspring, while minimizing its investment in the vehicles for the time being, offering minor updates or upgrades in search of capitalizing on the revenue derived from sales. By this time, Frey had left Ford and Iacocca had moved up to running the automaker and had many issues to solve, major among them was the growing significance of the imports.

Accordingly in light of the cancellation of the Half Cab in 1971, Ford partnered with Japan's Toyo Kogyo Company, which was better known as Mazda, and imported a version

of the manufacturer's B-Series mini-pickup6 to the United States as a cab chassis. It was shipped without the truck bed to avoid excessive taxation, which was known at the time as the "chicken tax,"[2] and assembled domestically. The small pickup was called the Courier, which had first graced Ford's lineup in 1952 as a sedan delivery based on Ford's two-door station wagon of the time. It was always marketed as a commercial vehicle and remained within the Blue Oval's offerings until 1960 when Ford began production of its Econoline cargo van. The Courier's front styling resembled the Ford's F-series including the grill and single headlights along with Ford badging that in subsequent years became much larger.

Nevertheless Bronco was certainly far from neglected. At the beginning of the model year, in September 1971, the Wagon had a new column shifter handle design that enabled owners to shift their "three-on-a-tree" more smoothly, and by mid-year, Ford introduced the Ranger trim package, in an attempt to portray this option as a higher-end choice for Bronco enthusiasts. The Ranger "bundle" consisted of new white striping on the body and hood, a unique silver grille, color-keyed front and rear pile carpet, deluxe wheel covers, wood-grained door trim panels, a Ranger-identified tire cover complete with bucking horse logo, cloth-inserted bucket seats, color-keyed instrument panel paint, and a fiberboard headliner.

With the discontinuation of the Half Cab by 1972, the Wagon was the sole offering of Bronco, but it still had some tricks up its sleeve with a new column shifter handle design and a new Ranger trim package, along with some additional interior updates. © 2023 Ford Motor Company.

1973

After seven years, a refresh was in order, but little was done to the shape of the boxy little sport utility. Generally manufacturers do some sort of model change by this point in a vehicle's life, and plans originally called for a new design by 1974. However, the oil crisis/embargo preempted any major attempt to introduce a revised vehicle, partly due to the resulting economic downturn. Instead the existing Bronco carried on by relying on cosmetic and convenience upgrades that stylists, marketers, and ultimately, accountants deemed more vital to Bronco's ongoing success than anything resembling a physical modification.

1973 Ford Bronco with Explorer package in Grabber Blue, one of two unique colors offered for this mid-year model. © 2023 Ford Motor Company.

Still, by this point, Ford's competitors had not been idly standing by. GM responded with the Chevrolet Blazer, and International Harvester brought out the revised Scout II. Jeep was firm in its belief that its CJ was strong enough to stand up to all comers and it soldiered on, relatively unaffected. But Bronco had to respond and introduced the C-4 automatic transmission and power steering as options. Additionally, the six-cylinder was increased to 200 cu in (3.3L), and the J-handle transfer case shifter was introduced—replacing the T-shift—shortly after the model year began, to the appreciation of many off-roaders.

Inside Bronco, orange—a color reflective of the era—was available for the first and only year, and most importantly, coming late in the season, the Explorer Package was released. It was primarily an appearance option, offering two unique colors, Grabber Blue and Burnt Orange, and a choice of matching "random-stripe pattern" cloth seats and badging on the glovebox. To further cement the identity, a special Explorer spare tire cover was included. It made for a nice appearance and definitely gave Ford dealers more choices.

Customer reaction to the Cruise-o-Matic was "automatic," especially among those who used their Bronco as a daily driver in urban areas. © 2023 Ford Motor Company.

1974

The list of changes was only a handful: a dome light was installed as was lighting within the automatic transmission selector. After the orange interior of the previous year, this time, a red interior was offered for the first and only time. The steering wheel was adapted to accommodate a thin vinyl horn button. At mid-year, electronic ignition was incorporated into the 302. California disallowed the 200 cu in (3.3L) six and enforced a new emissions package for all Broncos sold within the state.

Finally after roughly 500 Baja Broncos—though sources differ on the total amount, which makes the vehicles that much more desirable—the Bill Stroppe Baja Bronco program was discontinued at the end of the year.

The oil crisis of 1973 killed Ford's decision to refresh Bronco, but by 1974, minor upgrades were applied, many safety-related such as dome lights. A new red interior was offered, and by mid-year, electronic ignition was incorporated into the reliable 302. © 2023 Ford Motor Company.

1975

Stricter emissions were the big focus for every vehicle by mid-decade, and all vehicles were modified to accept unleaded fuel only along with the new catalytic converter. The F-Series steering wheel was a new upgrade to Sport and Ranger models. Sales dropped—this was not unexpected across all lines—but it was significant nevertheless, as only 13,125 Broncos were produced, based on demand. Were buyers starting to look for better gas mileage alternatives?

By mid-decade, every manufacturer was saddled with new emission rulings, and Ford and Bronco were part of that equation, as its engine was modified to accept only unleaded fuel while the new catalytic converter was installed. In an effort to spread investment costs, the F-Series steering wheel was installed in Sport and Ranger models, intended as an "improvement." © 2023 Ford Motor Company.

1976

Officially it was Bronco's 11th year of production, but 1976 "technically" marked 10 years since Bronco was introduced. It also was America's Bicentennial, and it was a time of celebration. Ford recognized the historic occasions by adding power-assisted front disc brakes and upgrading Bronco's rears to larger drums. The steering box ratio was further shortened to make maneuverability even more effortless. A front anti-sway bar was introduced, helping improve stability and decreasing rollover percentages. The Ranger Package had two special interior trim colors of Ginger and Avocado.

Ford also introduced a Special Décor Package late in the model year to create another series of distinctive Broncos, knowing there would be an audience for its application to the vehicle. Available only on the base Wagon and the Sport model, the package included a body-color roof, a color-contrasting stripe across the hood and upper fenders and down the flank, along with a blacked-out grille and matching headlamp rings. The kit gave Bronco a tough, "in-your-face" look that was attention-getting from its basic "from the factory" appearance.

Some exciting changes appeared on the 1976 Bronco, including power-assisted front discs and a shortened steering box ratio to make maneuverability easier. Ranger received special interior trim colors, and the regular Bronco wagon earned a Special Décor Package late in the model year to make it distinctive from previous years. © 2023 Ford Motor Company.

Stripes can make a difference or set a tone for any vehicle, even one that's already as distinctive as this 1976 Bronco. © 2023 Ford Motor Company.

1977

This was it—the final year of what would be known as the first generation. Certainly, Ford's Design Center was busy with Bronco's successor, its second gen, but there was still time to release this last version of what would become the pioneer of sport utility vehicles.

Most noteworthy from a design standpoint, the vehicle's filler caps were replaced with gas tank doors, finally giving Bronco a smooth plane along its flank, much appreciated by both stylists and purists. In conjunction with this, after years of shrinking the size of the auxiliary fuel tank (now just eight gallons [30 L]), the rear gas tank was redesigned and expanded in capacity from 12 gallons (45.4L) to 14.4 gallons (54.5L). And from a safety aspect, the front parking brake cable was better secured while rear market lights were now mounted vertically to give better clearance to the doors. From an enthusiast standpoint, the heavy-duty nine-inch (23 cm) rear-end housing was introduced. Interestingly the previous standard item padded instrument panel was now optional.

The Ranger Package for the final year of the "cube" Bronco was similar to the previous year, but the trim colors of Ginger and Avocado were now simply called Tan and Jade. The rear seat was required for the Ranger trim.

It was the last of the line, the end of the small Broncos for 43 years, a lifetime for many, but immortal to all, thanks to its enduring legacy.

The first-gen Bronco sold respectably at close to 20,000 copies a year for most of its run, and it helped plant a seed—the notion of a utility vehicle you want rather than need—that would truly take off after production ended in 1977.

Dave Kunz is Los Angeles KABC-TV's automotive reporter, and he knows a classic when he sees one. He's enjoyed this 1977 Bronco Ranger for many years. *Courtesy of Dave Kunz.*

SUMMARY: 1966–1977

NUMBER PRODUCED
225,797 including 5,000 roadsters, 17,262 pickups, and 203,535 wagons[3]

ORIGINAL PRICE
$2,404/$2,480/$2,625 (roadster/pickup/wagon, 1966), $5,078 (1976)

During those 11 years, Ford produced exactly 225,797 examples of the first-gen Bronco. Ford sold most Broncos in all configurations in its first production year, 23,776, though 1974 was its best year for the Wagon at 25,824, while 1975 was its worst year with just 13,125 vehicles delivered to customers. The modest production numbers, rust issues, and tough life that most of those Broncos had are the primary reasons why the first generation is rare nowadays and sought after by collectors.

---

[1] Zenlea, D. May 18, 2019. "The Original Ford Bronco Is a Vehicle Which Combines the Best of Both Worlds." *Automobile.* Retrieved 22 February 2023 from https://www.motortrend.com/vehicle-genres/1966-1977-ford-bronco-collectible-classic-review/

[2] The "chicken tax" stems from a form of protectionism that was used during a particularly perilous period of the Cold War by the United States under President Lyndon B. Johnson in response to tariffs placed by France and West Germany on importation of U.S. chicken. The president correspondingly imposed a 25 percent tax on the importation of potato starch, dextrin, brandy, and light trucks, which were designed to impact those two countries. Initially light trucks were added to stem the tide of Volkswagen Type 2 commercial pickups. However Japanese manufacturers also felt the sting from the sizable fees and pulled those models out of the U.S. market. Ironically companies did find loopholes, such as shipping trucks over sans truck beds or cargo boxes, but the ultimate result was the development of factories by these import manufacturers on U.S. soil, which today, number 21 and employ more than 150,000. The old saying stands, "If you can't beat them, join them."

[3] Source for Bronco sales figures provided by https://www.bronco-corral.com/articles/ford-bronco-production-numbers/

Nothing has quite the presence as a beautiful, tall,
second-gen Bronco in 1970s stripes. This staged photo from
Ford's ad agency, J. Walter Thompson, shows a 1979 Bronco
posed on the beach, demonstrating that the vehicle was
comfortable in any environment. © *2023 Ford Motor Company.*

Chapter 4

Bronco Generations Two through Five, 1978-1996

Some humans tend to get heavier as they get older, possibly due to genes, lack of exercise, or just too many desserts. What's the excuse for vehicles? Why do they often get bigger and heavier with each subsequent generation?

Technology, safety, and modern upgrades all play a role in expanding the size and weight of vehicles. Consumers also want more storage capacity, more doors, and more interior room. To cite just one example: The gas tank on a 2022 Bronco is either 16.9 or 20.8 gallons (63.9 or 78.7L), depending on whether you get a two- or four-door vehicle respectively. It was generally 14 gallons (53L) from 1966 to 1977. A gas tank on a new F-150 carries 23 gallons (87L). A 1966 F-100 had a 19.5-gallon (73.8L) tank. Bigger gas tanks mean more weight when fully loaded and more space needed to accommodate the bigger tank.

The point here is that the 1978 Bronco, the second generation, was indeed larger and heavier than its predecessor.

The trend among the Big Three (General Motors, Ford, and Chrysler) at this time was to downsize their vehicles for the sake of fuel economy and to compete with (primarily) Japanese imports. Many full-sized models were having inches lopped and weight cut in a resolute effort to produce lighter, more fuel-efficient vehicles. But when it came to trucks, all bets were off. The North American consumer market liked its trucks and SUVs big and tough and still does today. Even in the late 1970s, large pickups were unassailable. Ford sold 1,205,952 full-size trucks in 1978, including F-series, Broncos, Econoline vans, and Rancheros. Despite the panic from the second fuel crisis at that time, American know-how ruled the day.

ANNOUNCING THE
ALL-NEW '78 FORD BRONCO.

The first 4-wheeler that puts it all together.

1. Big Cube 351 V-8.
2. Part-time 4WD with optional automatic.
3. Rear footwell.
4. Rear flip-fold seat option.
5. Four-speed transmission.
6. Front quad shocks option.
7. Free Wheeling package option.
8. Off-road handling package option.
9. Privacy® Glass option.
10. Choice of bucket or optional front bench seats.
11. Front stabilizer bar.

COMPARE IT TO ANY 4-WHEELER, ANYWHERE.

And there it was. The all-new, much larger, much grander Bronco for 1978. Four-wheel drive vehicles were becoming plentiful on the dealer show floors, and Ford had to respond with something completely different. It did and sales took off—again. © 2023 Ford Motor Company.

This does not mean that growth was wrong. The little cab that represented the iconic first Bronco established a purpose, a lifestyle, and a category. For Ford and its dealer network, Bronco was a profit center, but also a customer connection. The goal for both was to maintain and expand that relationship, as well as build sales momentum.

And it was momentum that paid dividends for Ford. While the 1978 Bronco was larger, it offered a great deal more vehicle than the previous model and presented a modern appearance and capability. The new Bronco now matched the size of its competitors, their ranks enlarged with the entry of Dodge and Plymouth mounts to go along with

those from Chevy, GMC, Jeep, and International Harvester. What's most fascinating about the second-gen Bronco is that it lasted only two years. It was likely planned to be introduced earlier than it actually was, while its successor, the third gen was essentially ready to go by the time the second generation finally launched in 1978. In the course of those two years, the second-gen Bronco almost equaled the previous model's sales figures for the entire 11 years. The timing was obviously right. In its first year alone, Bronco attracted 78,000 customers. This was followed in 1979 with another 104,000 buyers. Those were serious numbers.

The reason?

First, in those years new models released every September. However, Bronco's second gen was introduced in mid-1977, several months earlier. The new Bronco was met with considerable approval and demand quickly exceeded supply, forcing some buyers to wait six months or more before they could get their SUV from a local Ford dealer. It was a plotline that would be repeated 44 years later.

Second, Ford's designers and engineers created the second-gen Bronco with serious off-road capability. Despite the onset of a second oil crisis amid other international fears, Americans needed an outlet, fanning the off-road craze, the increasing fascination with RVs, and the desire to explore the open road. All of this was fueled in part by the CB radio mania and such road movies as *Smokey and the Bandit* and *White Line Fever*. Bronco and other SUVs satisfied that growing urge.

Bronco's second gen was designed off the F-Series platform as early as 1972, and it assumed more truck-like characteristics and qualities, including increased length (a wheelbase 12 inches [31 cm] longer) and width and a weight gain of some 1,600 pounds (726 kg), all of which translated into a more solid ride and more interior space.

Despite the oil crisis, no six-cylinder was offered. Two V-8 engines handled the propulsion duties: the 351M, a two-barrel carbureted 351 cu in (5.8L) Modified V-8 and an optional 400 cu in (6.6L) V-8. Interestingly each offered similar horsepower, 156 hp and 158 hp (115 kW and 116kW), respectively (some sources list the output as 163 hp [120 kW] vs. 169 hp [124 kW]), but the 400 produced more torque. The new Bronco maintained its four-wheel drive, though a part-time system was standard. Also retained was the coil-sprung Dana 44 front axle and a leaf-sprung rear Ford nine-inch (23 cm) axle.

One factor that made the second-gen Bronco unique was its single-body style: a three-door wagon, or now, utility vehicle, with a lift-off rear hardtop attached behind the B-pillar. Bronco's revised tailgate introduced what would become a popular feature: a rear window that lowered into the drop-down gate.

Staying true to the Blue Oval family, Bronco's new face was common to the F-100, and it adopted features found within the long-running pickup, including air-conditioning, radio, and tilt steering. The wider interior also provided for a three-passenger bench seat in front, with a folding and removable rear seat, increasing capacity to six people, another first.

Finally Bronco employed new trim identities and model names. A Bronco Custom became the standard-trim model while Ford kept the Ranger name and selected the Ranger XLT as its top-of-the-line package. What set the '78 model apart was the fact that the Custom received round head-lamps while Ranger XLT had rectangular units that became standard for all Broncos in 1979 and beyond.

For both Bronco second gen's models, Ford offered what it called a Free Wheelin' cosmetic package. Capturing a sense of the "truckin'" lifestyle, the tri-color striping and blacked-out exterior grille further expressed the lifestyle of Bronco owners and the joy they bring.

THIRD GENERATION, 1980–1986

One of the third-gen Broncos that stood out early in this new wave of vehicles was designed for a single purpose, and one might say a higher power, if so inclined. That would have been the 1980 *Popemobile*, a specially modified Bronco that was used by Pope John Paul II during his October 1st through 7th visit to the United States in 1979. The Bronco was painted Wimbledon White, with an interior décor finished in Wedgewood Blue. This unique vehicle was configured to be open in the rear so the pope could safely stand and wave, or "greet" as the Ford press release maintained, his "friends and followers."

Upon completion, the special Bronco was delivered to the U.S. Secret Service, which supervised all travel and public exposure of the pope while in the United States, as it does with all foreign dignitaries, in addition to U.S. offi-cials, and of course, the President at all times. Following its brief use by the pope, the Bronco returned to Ford and was stored, with no subsequent information available. Would

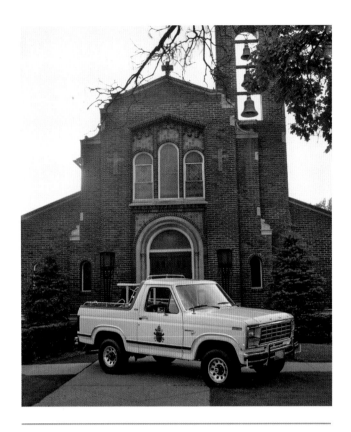

Designed to allow easy egress as the Pope addressed crowds throughout his 1979 visit to the United States, this modified 1980 Bronco proved extremely accommodating. © 2023 Ford Motor Company.

The Bronco *Popemobile* was for real, fashioned from a 1980 Bronco to accommodate the Pope during his 1979 visit to the United States. © 2023 Ford Motor Company.

From an exterior viewpoint, Bronco's new face was common to owners of the F-100, and the second gen's interior earned new features found within the long-running truck, including air-conditioning, radio, and tilt steering. © 2023 Ford Motor Company.

Vatican City have wanted it? Possibly, but officials there possessed their own automotive connections (and still do).

What's important about this third-generation Bronco is that it was much like trying on a new outfit and deciding upon a slightly more modern version. So, in a way, fashion played a role. The design work began in 1977, before Bronco second gen was offered for sale, but the new Bronco addressed some of the drawbacks and weaknesses its predecessor faced prior to its release and subsequent to it. Accordingly, the 1980 Bronco was shorter, lighter, and sported a more efficient powerplant—the return of a six-cylinder as the standard engine—but was naturally intended to be a full-size SUV.

Engines were a continuous variable, as emission standards and gas prices continued their overwhelming influence. Thus, the inline six, a whopping 4.9L (300 cu in), came with a four-speed manual transmission. The largest V-8 available was the 351M, as the 400 V-8 was dropped

1981 FORD BRONCO

Bronco Ranger XLT

Ranger XLT interior

Free Wheeling Bronco

(A)

(B)   (C)

Bronco is an advanced family-size 4-wheeler. It's the only one that offers Twin-Traction Beam independent front suspension (A). Twin-Traction Beam contributes to better off-road control and gives you a better ride than 1979 models. Bronco, trim and tough-looking outside, is family-size inside. There's room for six adults with front and rear bench seats. Choose from a list of options that includes Automatic Locking Hubs (B), and a special Snow Plow Preparation Package (C). For free-wheeling family fun, Ford Bronco is the one!

The blending with the F-Series was completed when Bronco entered its third gen in 1980. This 1981 model highlighted an attractive grille and recessed rectangular headlights and again demonstrated that it could go anywhere with its four-wheel-drive capability. © 2023 Ford Motor Company.

and the 302 V-8 returned as the standard V-8 engine. In 1982, the 351 Windsor debuted in the Bronco, replacing the 351M. A 210 hp (154 kW) "high output" version came two years later, and in 1985, the outstanding 302 (now referenced as the 5.0L [302 cu in]) lost its carburetor in favor of a multiport electronic fuel-injection system, increasing its horsepower output to 190 hp (142 kW), a considerable upgrade from previous years. Best of all, fuel economy was improved. A four-speed overdrive automatic transmission also became available in concert with the new engine.

The Bronco third-gen chassis was based on the F-Series, which was 500 pounds (227 kg) lighter to increase fuel economy. This new Bronco came across as more dignified than its older brother, more sophisticated, with a confident stance, an attractive grille, and recessed rectangular headlights. Bronco third gen retained its Dana 44 front axle with Ford's Twin Traction Beam independent front suspension. The transfer case was replaced with either a New Process 208 or Borg Warner 1345 version. The blending with the F-Series was completed when Bronco accepted the same trim levels, with Bronco (stand-alone) as the base, followed by Bronco XL, and Bronco XLT at the top. The first Eddie Bauer–based trim package for Bronco was added in 1985, featuring the now-familiar color-keyed two-tone exterior and branded identity within.

Updates over the years were mostly wrapped around appearance points: the Ford letters on the hood were removed and the Ford Blue Oval logo began to appear on the grille as well as attached to the tailgate in the lower left corner, increasing corporate identity.

Other styling changes included warning buzzers and an underhood light in 1983 and an optional 351 cu in (5.8L) Windsor V-8 with a four-barrel (4 BBL) carburetor in 1984 known as the 351 HO (high output) and rated at a respectable 210 hp (154 kW).

Though the new edition was a strong entry within the increasingly competitive sport utility lineup, Bronco's annual sales weren't able to reach those of the 1978–1979 model, but still did well, averaging approximately 43,000 annually for its first five years (actual first-year sales were 44,353), but then climbing to 54,562 and 62,127 for 1985 and 1986 respectively. Overall the seven-year production run released 331,306 units, according to Jamie Myler, Research Archivist at Ford Motor Company Archives.

FOURTH GENERATION, 1987–1991

A wind tunnel seemed to have played a significant role in the evolving design of Bronco, with the front end a product of aerodynamics including flush rectangular headlamps and wraparound turn signals and a smoother bumper application. Gone was any resemblance to the boxy cab of the original Bronco while the interior earned revised front seats, door panels, dashboard and controls, steering wheel, and instrument gauges.

Among the properties that were changed and upgraded within this generation was the initiation of electronic fuel injection (EFI) to both the 300 six and 351W 5.8L (351 cu in) V-8 along with rear anti-lock brakes when the new model was released. The new anti-lock braking system (ABS) was active for 2WD vehicles only, which boosted stopping power and safety. A five-speed manual transmission was introduced in 1988, and an all-new Borg Warner 1345 transfer case with optional Touch Drive electric shift replaced the New Process 208 case.

Sharing its appearance with the eight-generation F-150, though the chassis was retained from the previous model, the new Bronco was roughly the same length, width, and height as its predecessor. Wheel openings were more rounded in this version.

For the first time, two special editions were released during this series of production. Ford decided it was important to recognize Bronco's 25th anniversary, and in 1991, offered a Silver Anniversary Edition Bronco that was essentially a trim and appearance package featuring an exclusive Currant Red exterior paint and a gray, which others identified as "charcoal," leather interior (marking the first time leather was offered for seats in a Bronco).

The second model came out later in 1991, lasting only to 1992, and was called the Nite Edition. As its name declares, the special Bronco was painted Raven Black and boasted a blacked-out exterior identifying it as Nite. Bodyside tape stripes were available only in Aegean Blue and Azalea Pink as well as corresponding interiors. Like the 25th edition, both vehicles carried a V-8 and automatic transmission.

Overall Bronco fourth gen could be classified as a refinement more than an evolution. It took much of its looks from the F-Series, but it mostly cruised along, although it certainly gained valued upgrades. Ford should have made a bigger deal out of its 25th anniversary. It's surprising that

The fourth gen of Bronco began in 1997 when the SUV became more aerodynamic as headlights were flushed and turn signals were wrapped around the body, which featured an overall smoother appearance attributed to the bumper. Bronco was growing up and also included a 351 Windsor 5.8L (351 cu in) V-8. © 2023 Ford Motor Company.

The 1991 Silver Anniversary Bronco, in Currant Red only, was a trim and appearance package primarily, with appropriate badging indicating its significance in celebrating Bronco's 25th year of production. *Courtesy of the author.*

Another special model for the 1991 model year was the Nite Bronco, painted Raven Black with a blacked-out exterior. The only thing that didn't tie in with the dark theme concept was the body side stripe that was offered in either Aegean Blue or Azalea Pink. © 2023 Ford Motor Company.

automakers didn't seem to think that was important in the 1990s when such milestones were hitting vehicles launched in the 1960s. Nostalgia hadn't begun playing such an important role in people's lives yet, at least not to the point it is today where there are so many individuals, particularly baby boomers, who long for memories of their youth. Commemorating any such big event today is an opportunity to unveil a special trim package, at least, tied to social media, merchandising, product placement, influencers. . . .

So, Bronco continued. It sold well, but within Ford, attention was definitely being paid to the next SUV being prepped for launch.

FIFTH GENERATION, 1992–1996

When Bronco fifth gen was introduced in September 1991, no one knew for sure how long the big SUV would last. The Explorer had arrived in March 1990; its original intention was to replace Bronco II. But almost immediately, Explorer's ability—particularly with its four doors—to extend beyond the 4×4 buyer and intrigue families who wanted something different from the minivan or station wagon quickly led to its sales success—and spoke volumes that a new category of vehicle was about to be unleashed.

Explorer was being built in Lexington alongside its stablemate (and design source) Ranger but was in such demand that production of Ranger was held up to accommodate more assembly time to Explorer. The writing was on the factory wall.

That didn't stop Bronco designers from pulling out the stops on this fifth gen, regardless of its projected longevity. However, since it had been effective for much of its existence, this latest Bronco fifth gen was based on the newest generation—its ninth at that point—of the F-150. This time, stylists really got the looks right. Bronco fifth gen came across as polished and highly refined, as if years of touch-ups and reshaping had finally resulted in its finest expression.

Very little advertising supported the new generation's launch. The investment was in the vehicle, not so much in its marketing. The promotional dollars were spent on Explorer. Nevertheless, much was made in public relations efforts about the fifth-gen Bronco being an intelligent vehicle as well as classy.

The new SUV came with a driver's side airbag and a three-point seat seatbelt. Front bucket seats seemed even more comfortable, especially with optional leather seating and captain's chairs that were available in the XLT, Nite (which continued from the previous model), and Eddie Bauer edition. Captain's chairs also embraced power lumbar support. The center console could accommodate more, and the lip in front of the twin cupholders was taller, ensuring that a Big Gulp would receive adequate accommodation.

Elsewhere in the interior, the focus was on luxury and convenience with new gauges, the first-ever digital odometer, advanced climate controls, and premium sound systems. Adding to this list was the auto-dimming rearview mirror, illuminated visor mirrors, an overhead console (in 1994), rear cargo net, and redesigned power window and door lock positions. Optional keyless entry and an anti-theft alarm also were available. This was *definitely* uptown.

Externally, like the F-150, the SUV was rounded at the front. Its headlights were more comfortable now as they wrapped around the corners, and the grille more traditional—conservative and solid. Additionally, a herd of new colors was offered with a number of two-tone combos, Eddie Bauer flavors, and monochromatic schemes (black, red, or white) on the XLT Sport that included bumpers and grilles. The XLT level also offered a two-tone light teal green and white that was limited in number.

The biggest factor that never left was Bronco's ability to go four-wheeling. It still had the means and the desire to be pushed off-road. The 4.9L (300 cu in) inline-six, 5.0L (302 cu in) V-8, and 5.8L (351 cu in) HO V-8 (a.k.a. Windsor) remained, though the six-cylinder was dropped after 1993. This had always been Bronco's claim to fame, and it remained solid to the last unit on the assembly line.

Make no mistake, the SUV market was changing rapidly, within Ford as well as the other members of the Big Three and all importers. The tough off-road Bronco was being supplanted by a softer vehicle that was both more urbane and urban. Sales dropped, and for any vehicle "long in the tooth," this was a sign that it was "sunsetting."

Bronco fifth gen was modern, and it had luxury features. It had the comfort of a passenger car. But it was tall and still rode a bit rough thanks to its live rear axle. Yet, consumers wanted all the things Bronco offered but with four doors and a cushy ride. Even today, the importance placed on being comfortable in a vehicle cannot be underestimated.

Nevertheless, Ford continued updates throughout Bronco fifth-gen's lifetime, answering the call to make

ABOVE AND LEFT: Even though more attention was paid to the new Ford SUVs on the block, the Explorer and the upcoming Expedition, there was definitely a demand both inside the Blue Oval as well as among its most loyal customers to continue refinements of the existing Bronco model, now in its fifth gen, and to make it as desirable—and luxurious—as buyers wanted. Both this 1993 XLT and its later stablemate, a 1995 model, can say that the last of these gen Broncos went out in style. © 2023 Ford Motor Company.

the SUV more emission-compliant by dropping the practice of marketing the lift-off hardtop as removable. More rust-resistant bolts were installed, the mandatory Center High Mount Stop Lamp (CHMSL) or the "third brake light" was placed above the rear window on the hardtop, and different paint and upgrades for the Eddie Bauer edition were continued.

The final big salute to Bronco fifth gen obviously came in the form of the police parade led by Al Cowlings' white 1993 in June 1994. Within the rear lay O.J. Simpson. The story has been relayed elsewhere, but it's been written numerous times that some 95 million viewers watched that brave Bronco pace the I-5 Freeway and into history. For a brief period afterward, Bronco was on everyone's mind.

Bronco was on fire for a long time. It was, and still is, an incredible vehicle. Think about the number sold over each generation:

- First generation: 225,797 original Broncos were purchased over 11 years.
- Second generation: Despite thriving for only two years, 181,955 were snapped up.
- Third generation: 328,063 units sold in seven years.
- Fourth generation: From 1987 to 1991, Ford delivered 235,451 vehicles.
- Fifth generation: 162,703 sold in five years.

RIP VAN WINKLE + FIVE

Alas, all good things must come to an end, even if temporarily . . . at least a quarter century, anyway.

Bronco was quietly discontinued on June 12, 1996. The last model rolled down the assembly line to applause and tears, signifying the end of an era at the Michigan Truck Plant. What a reputation it had earned, and what a legacy it left. The workers knew their jobs were safe. After retooling during the summer, they were set to build the Expedition. Under development since 1991, fully approved by 1993, and with an investment of $1.3 billion, what was then known in-house as UN93 was ready to be for the assembly line. Ford's personnel knew what they needed to do and based on the experiences with Bronco, were ahead of schedule when it came time to release the new four-door, full-size SUV.

But Bronco would be back, after a long sleep. And its timing couldn't have been better.

THE LAST FORD BRONCO TO ROLL OFF THE LINE AT MICHIGAN TRUCK PLANT
JUNE 12, 1996

photo by Harold Bailey

Last off the line for (approximately) 25 years. Vehicle number 1,152,890, according to the Michigan Truck Plant on June 12, 1996. © *2023 Ford Motor Company.*

Had the police chase of O. J. Simpson in the white Bronco been a TV show, it likely would have been one of highest rated programs ever on television. The big question: Did media coverage of the Bronco increase sales during the SUV's last two years? Broncos did experience a growth spurt in 1995. *Alamy Stock Photo.*

The Dune Duster started its life as a Roadster and then was treated to a series of modifications by George Barris Kustom of Hollywood, California, the same man and company who created the dynamic vehicles on TV's *Batman*, *The Munsters*, *The Green Hornet*, and many others. © 2023 Ford Motor Company.

Chapter 5

Future Plans Require PROTOTYPES

O ne fascinating footnote to the history of Bronco is the fact that Ford stylists, designers, and engineers were always "blazing the trail" in terms of what else they could do with the marque, such as what other interesting looks and features could they collectively produce to stretch not only the longevity of the brand but also to generate persistent attention and test new equipment and applications.

The results were varied and limited at best. But the few times the staffs went out on a "public" limb in terms of legitimizing their efforts with photo shoots and corresponding press information, particularly as "auto show fodder," indicated that they were thinking of how Bronco could be driven by a slightly different audience or pushed harder toward a more functional outdoor or performance vehicle.

Whatever the strategy, the efforts were not in vain, as they successfully kept the creatives thinking, wondering what more they could do with Bronco in the years ahead. As the world knows, the best plan was to simply evolve the vehicle, not revolutionize it, but these examples below demonstrate that a few were wondering how much of an evolution they could get away with.

DUNE DUSTER

Not long after Bronco first gen's launch, plans to feature a special custom version at the 1966 New York International Auto Show April 9–17 began. The new Roadster model to Ford's lineup of vehicles was mocked up in the manufacturer's Styling Center in Dearborn with custom work carried out by George Barris Kustom in Hollywood, California.

Barris, of course, was well known at the time for his legendary TV vehicles including the *Batmobile* and the *Munster Koach* from the popular mid-1960s shows *Batman* and *The Munsters*, as well as the Torino from *The Green Hornet* and the Monkees' *Monkee Mobile*. He would go on to build many other cars for other popular programs and films. His Hollywood shop remained in its same location until 2021.

This modified Bronco was painted a "specially formulated Golden Saddle Pearl," stated the supporting Ford press release, complemented by walnut appliqué on the rear side panels and exposed chrome side exhaust pipes. Highlights included chrome wheel lip moldings accenting the machined steel alloy wheels with knockoff hubs, a redesigned hood incorporating a real air scoop for added engine cooling, suede padding wrapped around the dashboard, beige loop carpeting, and a tonneau cover for rear compartment protection.

THE WILDFLOWER

According to the modified press release issued by Ford on January 18, 1971, the Wildflower Bronco was a specially customized version that was "sure to be one of the most colorful show cars on display at automobile shows this year."

Created by Ford's Design Center, Wildflower was extensively modified in its previous existence as the Dune Duster show car five years earlier. The principal modification this go-round was in fact the paint, completed to "achieve a lively, carefree appearance with added luxury and safety features," which are highlighted by the "psychedelic design of blues, yellows, and reds [is] topped off by a pink grille." The design seemed like a day late and a dollar short. Weren't psychedelic colors bigger in the 1960s?

The roll bar, now color-matched to the rest of Wildflower, chrome side exhaust pipes, the wheels with knockoff hubs, and hood were all carry-overs of Dune Duster. Reflecting the style of the era were brightly flowered vinyl-upholstered bucket seats, matching instrument panel, and red loop carpeting throughout. To say it was wild was an understatement—the paint job at least. Janis Joplin would have approved.

Painted a "specially formulated Golden Saddle Pearl," the interior featured a suede padding wrapped around the dashboard with similar suede material on the seats. The beige hoop held vinyl padded headrests, while a tonneau cover provided protection for the rear of the vehicle. © 2023 Ford Motor Company.

Though the Roadster, on which the 1971 Wildflower was based, was discontinued two model years later, the main focus of this styling exercise was to highlight the "wild" paint and the bright flowers coating the vinyl-upholstered bucket seats. The theme was all about "a lively, carefree appearance." © 2023 Ford Motor Company.

BOSS BRONCO 1969

The Boss Bronco was based on the success of the high performance Boss 302 Mustang and its even more powerful brethren, the Boss 429. The former was built to compete in the Sports Car Club of America's (SCCA) Trans-Am racing series, and the 429 version was created to homologate its engine for NASCAR racing. In any event, these highly potent "pony" cars were well-received by the high-performance audience to which they were targeted and led Ford, particularly newly hired President Semon "Bunkie" Knudsen, known for his aggressive performance chops while at competitor Pontiac, to request a similar treatment to the company's popular Bronco.

Built through a collaboration with Kar-Kraft and Bill Stroppe in 1969, the Boss Bronco, referred to originally as the Special Bronco, featured a blueprinted GT 350 Shelby engine along with a modified Cougar Eliminator hood scoop to accommodate that big Windsor motor, as well as a Hi-Po C4 automatic, and was built to exacting standards as all Kar-Kraft's vehicles received. Ford was Kar-Kraft's only client and exclusive special vehicle manufacturer. Its responsibility was to serve as both a race shop and custom producer—a "skunks works–creating Boss 302s, 429s, and most sacred of all, the GT40s.

If Ford's Mustang could have a Boss 302 and later a Boss 429, why couldn't there be a Boss Bronco? Ford contractors Kar-Kraft and Bill Stroppe thought it was a great idea, but sadly, Ford's management did not. *Courtesy of* Hot Rod *magazine.*

Painted yellow, allegedly Knudson's favorite color, sporting a white top with the same hockey stick-styled stripe as the Boss 302s running along the side, and complete with *BOSS BRONCO* lettering on the fenders, the intended goal was to offer a top-of-the-line high performance race-like truck that many Ford enthusiasts would want. Stroppe, of course, was already legendary with his off-road-winning Broncos, and his association helped validate the concept.

Unfortunately, Knudson was a victim of office politics, and being an outsider, having come from GM, he had few allies. Henry Ford II finally recognized this and let him go in September 1969. Upon his departure, the Boss Bronco went out the door with him. Ironically the single prototype built disappeared for 45 years, believed to have been acquired by someone internally, and only received recognition around the time the new Bronco was announced. Colin Comer of Colin's Classic Automobiles in Milwaukee, Wisconsin, is the present owner.

Embracing both a styling as well as a high-performance package, in retrospect, it was clear that Ford's stylists and performance experts were far ahead of their time as the Boss Bronco was quite similar to today's Bronco Raptor.

Knudsen was a survivor and remained in the automotive industry, serving as president of truck manufacturer White Motor Company and as a National Commissioner for NASCAR until his death in 1998 at age 85.

1972–73 SHORTHORN

The Shorthorn is one of those stories from within the Glass House (or the Product Development Center) that just will not settle down. Built in response to the success of the Chevrolet K5 Blazer, the design was based on the F-Series chassis. Dubbed "Project Shorthorn," a number of bigger versions were proposed, including Midhorn, Longhorn, and Widehorn, but the Shorthorn seemed the logical pathway—until it wasn't.

It has always been maintained that the first oil crisis of 1973 prompted Ford to hold off approval until gas lines got shorter and executives' willingness to take risks materialized. Turned out, the former happened sooner than the latter, and Shorthorn never got the green light it needed until a more modified version that became Bronco second gen debuted in 1978.

So the story goes, the Shorthorn was supposed to be the next gen of Bronco, based directly on sibling F-Series. However, the 1973 energy crisis that loomed its ugly head over transportation in general and vehicle sales in particular deemed that this prototype would be scuttled and the first gen would live on for a few more years. *© 2023 Ford Motor Company.*

A proposed 1980 Olympic Edition Bronco. Ford was the "Official Car of the 1980 Winter Olympics," which may have entitled the manufacturer to feature its trucks and SUVs. No doubt this Bronco was intended to bear the prestigious five-ring Olympic emblem and wreath. However, all things Olympic ground to a halt in March 1980 when U.S. President Jimmy Carter announced the United States would boycott the Russian-sponsored Summer Games due to the Soviet Union's invasion of Afghanistan in 1979. This Olympic Bronco quietly slipped back into the shadows as a result. *© 2023 Ford Motor Company.*

1980 BRONCO MONTANA LOBO

One thing to say about designers is that they are futurists. They know what is going to happen years, often decades, ahead of everyone else. They either do their research or simply have the kind of sixth and seventh sense that great detectives and entrepreneurs have when it comes to creating ideas and concepts that go far beyond today's standards and encompass the needs of the world down the road.

This is particularly evident as Bronco began to enter its third decade of existence and third generation of style and opportunity.

Choosing to maintain its two-door structure but adopting a combination of the original roadster cut though modified with a futuristic plexiglass door and B-pillar with trailing louvers, the Montana Lobo concept favored a long sloping framing to the end of tail in a wedge-shape and open bed, making it equal parts utility and pickup. A glass door separated the passenger space from the hauling space, the latter cargo area lined with hemp and included two fold-down bench seats. A tailgate retractable loading ramp folded into three sections.

Reflective of its name, the vehicle was built tall, wore off-road tires, and had flared wheel arches. Ford's 5.0L (302 cu in) V-8 fed exhaust to huge, exposed Corvette-like side pipes located just above extended running boards, and it featured massive front and rear bumpers. Large roof-mounted track lights completed the lifestyle-oriented package, similar to what most trucks favor today. The Montana Lobo was a favorite at the 1981 Chicago Auto Show when it was introduced, but that was it. Credit for the vehicle was due to Ford's Advanced International Design Center and the Ghia Construction Studios in Turin, Italy.

Taking on the appearance of a pickup, the rear of the Bronco Montana featured a conventional pickup bed with a glass door between the bed and the cabin. A small tailgate was in place, and presumably, the large rollbars had a purpose to balance weight and prevent rollovers. But overall, it was an odd silhouette, though it did have its moment at the 1981 Chicago Auto Show, one of the largest shows in the United States in terms of attendance. © 2023 Ford Motor Company.

1990 DM 1 CONCEPT

The Chicago Auto Show is the largest auto show in the United States, traditionally attracting more people than any other major-league auto show all season. That's why Ford chose this 10-day event in 1988 to launch its most compelling SUV interpretation of the Bronco thus far.

The DM 1 started out as a competition among industrial art students sponsored by Ford. Based on a Ford Escort platform, Derek Millsap's concept won out, highlighted by rounded edges and a large rear hatch that extended into the roof. Named after Millsap, the five-seat SUV featured a rounded body of steel-reinforced fiberglass radiating a small crossover-like appearance. It clearly had a Ford presence to it, and the automaker claimed the DM 1 could become four-wheel-drive, particularly had it assumed the Bronco mantle. Its presence in Chicago gained much interest at the time, but sadly, while ahead of its time, DM 1 did not get further. Millsap, however, was quickly hired by Ford where he spent 19 years as a designer/sculptor before moving on to another manufacturer.

1992 BOSS BRONCO (NUMBER TWO)

This is the second round for this name, now released for a prototype in 1992 and interestingly, or coincidently, painted bright yellow (with all due respect to Mr. Knudsen). While this was labeled a Bronco, it looked more like a Chevrolet Blazer S10. What makes this vehicle incredibly unique and very prescient for its day is the sloped rear (similar to the contemporary BMW X6 and Ford Edge). Visibility through the rear must have been peerless with a large pull-down glass and a short tailgate.

This four-wheel-drive concept was shown at the 1992 Chicago Auto Show and billed as having "comfort and convenience of a car," signaling the public's enthusiasm for a potential CUV. Massive horizontal headlights tucked within the body-colored grille and a nice, elevated hood could have logically accommodated a tidy little V-8 that would have made this the kind of Bronco enthusiasts would want. But then Ford launched the SVT Lightning to take that steam and noise, and the second Boss was retired.

Looking more like a Moon Rover than a Bronco, the DM-1. It was named for future Ford modeler/sculptor Derek Millsap, who designed this winning entry for a national collegiate contest sponsored by Ford. The prototype offered a new perspective on what a four-wheel drive vehicle might look like in the 1990s. Millsap enjoyed a long, successful career at Ford. © *2023 Ford Motor Company.*

This Ford Bronco Prototype dubbed Boss Bronco (second go-round) had a lot of Chevrolet Blazer in it or at least what a Blazer would look like at some point. The severe rake at the rear was visually appealing, though it did reduce storage capacity. Still, save for the roof-mounted lightbar, this design might have made for a very slick Bronco had it seen the light of day. © *2023 Ford Motor Company.*

1994 OUTBACK BRONCO CONCEPT

The design team was still at it, even though by now it was known internally that Bronco was done in two years. However, that didn't mean if Ford execs liked something from the design studio it couldn't be put into production as a limited edition or final-year option. Thus, had the timing been better, and playing to the increasing desire for aftermarket accessories and the call of the wild, the Outback Bronco, as proposed, could have become a reality.

Outback carried over some characteristics from Boss Bronco released just two years earlier, with a rounded C-pillar, though nowhere near as angled as the latter. Whereas the Boss would not allow a customer to remove the rear cap, it appeared that on Outback that may have been possible. And why not? This vehicle was designed to enjoy the open spaces, the outback, if you will.

With a cattleguard, large fog lamps, and a significant winch built in, protecting a billet grille, common among the aftermarket crowd at that time, the front end looked aggressive and ready to forge ahead. The same hood from Boss appeared on Outback, meaning the 5.0L (302 cu in) was the engine of choice, and having more horsepower was never wrong. Aggressive side pipes returned, sitting above an aluminum step, and flared wheel wells were smoothly incorporated into the body. It was a pity larger tires were not included. That's certainly something the new models have gotten right. Ride height was significant, meaning this SUV was prepared to do its share of off-roading.

The red-and-white color scheme with slanted red hashes was striking, and the Bronco logos sitting on the fenders were standard. At the rear, however, the bumper was removed, and the curved glass provided a more streamlined pathway to the tailgate, which was emblazoned with *Outback Bronco* across its width. Overall, it was an interesting vehicle, and for a select group of adventurists, as well as the government service fleets (fire, park, geological, etc.), it might have been a desirable choice.

2001 U260

Three years after Bronco left dealer showrooms, one team at Ford was still devoted to the legacy it held. Although it was a small group, it was fairly senior and influential in an underground, clandestine sort of way, which is why it was known—as the Underground. With the dawning of the new century, developing prototypes that placed Bronco at the foundation was an important piece of both design therapy and their perceived future of Ford.

In 1995 Ford brought in Moray Callum, a noted designer from Ghia, an Italian automobile design and coachbuilding firm owned by the Blue Oval, who would become Vice President of Design at the company. In 1999, however,

Not much change here, short of the rounded C-pillar and lift kit. This Outback Bronco Concept might have done well in foreign markets. © *2023 Ford Motor Company.*

Early version of U260 displayed outside for comparisons of grilles and trim pieces. © *2023 Ford Motor Company.*

The prototype the public never truly got to see until years after it had been built started its life like every other design—on a clean sheet of paper. The beefy SUV took some references from other manufacturers, but retained true Bronco characteristics. © 2023 Ford Motor Company.

Callum led the Underground in developing an updated version of Bronco, with the precept of maintaining its overall look and structure along the lines of the original—an uncomplicated, easy-going off-road vehicle, but moving away from the full-size character of its more recent ancestors.

Code-named U260, with the *U* for Utility, *2* for two-door, and *60* for the Ranger's T-6 platform on which the new Bronco would be based, the Bronco never got a headwind due to the automaker's ongoing battles and financial setbacks from the Firestone tire/Explorer rollover misfortune that seemingly ground all non-revenue-generating projects to a halt. The single prototype was stuffed in a corner and only saw daylight when the vehicle was publicly revealed for the first time in the lead-up to Bronco sixth gen's release at the 2021 annual Woodward Dream Cruise in Ferndale, Michigan, in suburban Detroit.

2004 BRONCO CONCEPT

No better person than nostalgia king J Mays, designer of the successful New Beetle, took command of the 2004 Bronco, launched to great acclaim at the January 2004 North American International Auto Show, a.k.a. Detroit Auto Show. At the time, Mays was Ford's Group Vice President of Design.

Pulling out onto the stage within Cobo Hall that Ford claimed every year, the Bronco prototype, featuring a monotone color scheme expressed by its stand-out silver finish, inspired nods of approval from the typically cynical media in attendance. With Mays serving as host, the stripped-down concept bore much similarity to Bronco first gen, something Ford's accompanying press material called "demonstrating Bronco's authentic spirit while advancing today's powertrain technologies."

The signature three horizontal bars found on the F-Series were there with the Bronco nameplate wedged directly into the middle (not too dissimilar from today) and so too were the traditional round headlights and boxy, squared-off look, which pleased the crowd and the purists. Tied to that were exposed door hinges, cowl vents, and flared wheel wells.

"True to its heritage, the Bronco concept is a tough, genuine SUV that's all about function," said Mays, (he retired from Ford in 2013).

To underscore the concept's functionality, it was explained that the vehicle carried a 2.0L (122 cu in) inter-cooled turbodiesel with an efficient six-speed PowerShift transmission and four-wheel-drive system. Supporting Mays, additional executives indicated how Bronco could easily complement its existing SUV lineup.

"The Bronco concept strikes a familiar profile of the authentic SUVs of the late 1960s and at the same time is contemporary, appealing and relevant for today's market," Mays added. "It's like your favorite pair of worn, faded jeans—classic, familiar, comfortable and always in style."

While that may be true about one's jeans, despite the concept's strong reception, sadly, it remained in the closet and didn't get a chance to be anyone's favorite.

2013 EXPEDITION "SPECIAL EDITION" CONCEPT

The United States Patent and Trademark Office (USPTO) is responsible for all trademarks and patents registered in the United States. If one were successful in receiving approval for a trademark, for example, that mark is thus protected but the applicant of said mark does not own the word or an image but the right to use it to identify the *source*, which is the owner of a good or service. Trademarks do not expire after a set period of time like patents or copyrights; however, if the trademark is not being used, the USPTO deems it should not be protected. Trademarks are not meant to be stockpiled for future names or logos, for example.

Thus, trademark holders are required to prove to the USPTO every five years that their mark is still in use. The 10th year requires actual proof that the trademark is in current existence, which includes "photographic evidence of a product, using the trademark, is available for sale." This process is repeated every 10 years.

Accordingly, when Bronco ended production in 1996, this meant that by 2006, Ford had to show the name was still being used. (The author was involved with this same process when employed by Nissan North America and the Nissan Z had been briefly discontinued in the United States in the late 1990s.)

With significant amounts of leftover dealer inventory, warranty fulfillment and the like, Ford could show the mark was still in use for several years after that, but by 2013, Mark Grueber was one of the few at Ford who knew the Bronco mark was now in trouble after years of nonuse. With support from the Underground, he created a special "show" Expedition that brandished a Bronco badge and displayed it as a show vehicle in the Ford booth at the Specialty Equipment Market Association (SEMA) Show. No one went crazy, which was surprising, but the task achieved the effect. The trademark was in use. Case closed. The next go around wouldn't matter as the formal announcement of the new Bronco came in 2017.

2020 BRONCO R PROTOTYPE

See Chapter 13 "Racing Becomes Reality." Although technically, this racing Bronco was a prototype, it was built as a racing machine, actually more as a test vehicle in a motorsports environment, but it will be briefly covered there.

The Bronco prototype that so many purists hoped would be built. The 2004 Bronco Concept unveiled at the North American International Auto Show represented everything the original Bronco first gen possessed, but Ford just wasn't ready . . . yet. © *2023 Ford Motor Company.*

The 2004 Bronco Concept debuted at that January's
North American International Auto Show in Detroit and

Chapter 6

The Rebirth of
THE BRONCO

Of course, the first question with respect to Bronco's rebirth is, "What took so long?" There are a lot of answers, but when one considers Bronco's place in Ford's history, it really was a considerable amount of time that elapsed after the last (fifth) generation ended on that sad day of June 12, 1996—seemingly forever.

But there is rarely a "forever" in the auto industry. Well, it's hard to imagine the Edsel nameplate ever resurfacing, ditto the Pinto, but it's amazing how many models experience a resurgence when the timing is right. Consider Ford's Maverick nameplate. It was definitely not the comeback one expected. Its revival as a lightweight truck again shows Ford's knowledge of the marketplace, and that savvy has paid off with a nice response by customers.

Clearly, it's more than just celebrities and politicians who get a second wind.

Bronco returned with a vengeance. Certainly, Ford toyed with the idea for many years. Witness the 2004 Bronco Concept that graced the Cobo Hall stage during the annual North American International Auto Show. For those in attendance, including the author, the vehicle looked fantastic, and calls for Job 1, what Ford calls the first batch of a new vehicle, were immediate.

But it never materialized, and depression set back in among the devoted.

Then, in the mid-teens of the twenty-first century, there were more rumblings from within the Design Center. Magazines wondered "what if?" and offered their own digital versions of what a new Bronco might look like.

These speculations bore many similarities to the 1996 model, a sort of skinny Explorer, an F-150 façade. Ford fanned the flames. Enthusiasts continued to believe a new model "could" happen.

MIRACLES DO HAPPEN

When Ford announced in 2017 that it was going to give the Bronco another run, enthusiasts, journalists, and off-roaders everywhere were ecstatic. Competitors were not. One thing was certain. It would have circular headlights and a traditional look. Calendars were set and days were ticked off waiting for a firm(er) launch date. Plans at that time called for a 2021 release.

But then, the COVID-19 pandemic and a series of unfortunate setbacks all hit simultaneously. The Bronco went to an uncertain, some say "soft," time frame. It was all relative to those who had already been waiting for years.

Yet, positive thinking can work. After multiple sighting reports, faithfully recorded by every automotive-related journal in the world, online posts, and thousands of opinions expressed among both the well- and misinformed, spread within the enthusiast community and on dedicated forums, Ford finally announced on June 13, 2021, that the 2021 Bronco would be revealed on July 9, 2021.

Jump ahead six days and, hold on, Ford issued another release stating that the new Bronco reveal date would move to July 13th.

WHAT? WHY?

Purists quickly knew why. July 9th is a day most Ford executives, both past and present, and a lot of other Bronco fans would just as soon forget: O. J. Simpson's birthday. But we'll get back to him in Chapter 10.

Right, let's just move that reveal date up a few days to the 13th. No big deal. Nothing to see here. That July 9th choice was "purely an unintended coincidence," Ford publicists noted/winked. Case closed. This was, after all, the third rescheduling for the reveal, thanks to a cancelled auto show and private unveiling, due to the ongoing pandemic

The new Bronco debut would be digital, so Ford marketing pros decided to partner with Disney. Why not? Is Disney not the home of Marvel? Star Wars? The Magic Kingdom? Where dreams are made? Its creative agency, Disney CreativeWorks, would develop a troika of three-minute videos to run on Disney-owned properties ABC, ESPN, and National Geographic the evening of the 13th of July on *prime time*.

The "films," as they were referenced, all featured someone who logically had a tie to the perceived Bronco-buying public: professional climber Brooke Raboutou was the focal

A number of "what-ifs" appeared following Ford's 2017 announcement that the Bronco would return. Lacking definitive information, more imaginative fans began to create digital speculations of what a new Bronco might look like as seen in this clever interpretation of a how an Explorer might morph into an attractive two-door Bronco. *Courtesy of Bronco6g.com.*

The Bronco-reveal timing was well-planned. More than a TV commercial, it was story-telling at its best with short videos featuring country music singer Kip Moore, professional rock climber Brooke Raboutou, and Oscar-winning director Jimmy Chin, known for his rock-climbing photography on ABC, ESPN, and the National Geographic channel. It was Ford's strategy to focus on what the Bronco lifestyle could offer owners of the new vehicle, particularly those who enjoy playing outdoors. © *2023 Ford Motor Company.*

point for the spot that ran during ESPN's "Sports Center," while Country Music Singer Kip Moore appeared in a second spot placed within the "CMA (County Music Awards) Best of the Fest" on ABC, and Academy Award-winning Director Jimmy Chin (a 2019 Best Documentary Oscar for *Free Solo*), who is also a professional climber, was spotlighted in National Geographic's "National Parks: Yosemite."

Naturally, the costar to each of these films was a Bronco. And with Disney planning at the helm, the videos played well, and everything went off without a hitch. Viewers, and most importantly potential customers, were driven to Ford. com where they could find more information and place a $100 refundable reservation for their very own Bronco.

THE LAUNCH

More than 13 million people tuned in either on television or watched the unveiling on computers or other mobile platforms. Within hours of the July 13th reveal, #FordBronco was the dominant hashtag throughout social media. According to Google, more than 5.4 million visitors landed on the Ford.com website within 48 hours. By the following day, all of the 2,000 limited-run Bronco Sport First Edition models were ordered. The reaction was the same for the bigger Bronco two- and four-door First Editions.

Within three weeks, the number of deposits hit 165,000. That translated into a nice little $16,500,000 addition, enough to put a small dent into that Bronco marketing

Both Bronco and Bronco Sport offered buyers a First Edition in their inaugural year, making 7,000 customers per vehicle very happy by purchasing what will become a relatively rare model in years to come. © *2023 Ford Motor Company.*

Though Ford offered many different versions of its Bronco, from mild to wild–Wildtrak, in this case–to the Raptor, the designers, engineers, and marketers at the Blue Oval realized an important aspect of personal ownership is the ability to customize one's own vehicle, for which the sixth gen Bronco has been enabled, with upgrades available from Ford Performance parts. © 2023 Ford Motor Company.

One customer connection that sets Bronco apart is the Ford Bronco Off-Roadeo. It's an annual off-roading program, a one-day "school," that allows owners of a new Bronco, at no cost to them, get a feel for the SUV's performance on an "outdoor adventure playground." It also allows Ford personnel to directly interact with their buyers and gain valuable feedback on everyone's favorite vehicle. © 2023 Ford Motor Company.

budget. Of course, that money was all refundable, and the wait was going to be long (and even longer as it transpired), but if you had been waiting for more than 20 years already, what was another 20 months?

Ford's Media Center made sure to turn the heat up high on hype too. Celebrating the Bronco's 55th birthday in August 2020, Ford recognized the "unprecedented response" to both Bronco and Bronco Sport, which it called "Ford's new outdoor brand," in an attempt to tie a whole lifestyle vibe to the vehicle. Bundled within the release were the additional announcements of a return to the Baja 1000 off-road race where Ford would once again campaign a Bronco, the introduction of a new hands-on consumer driving experience called Bronco Off-Roadeo, and the first of what would become many invitations to potential buyers or people already within the reservation line to personalize their ride with five "adventure-inspired concepts."

Thus was born the capacity for individual lineups and the commencement of a brilliant menu of models, as well as options, that enabled every buyer to customize their Bronco depending on packages and of course, their wallets. It was nothing new, of course, but the magnitude of the offerings was significant and its presentation captivating. The order banks continued to swell.

Ford's Bronco Marketing Team was on the job too, with a new tagline for the brand. Broncos would be "Built Wild." Technically they would be built at the Michigan Assembly Plant, which is more appropriately called "controlled chaos," though overseen by a very cool, high-pressure, well-organized team. Quality remains Job 1, after all.

Built Wild became a promise that the renewed Bronco brand was going to be bold and adventurous, that going off-road was intended to be as common as running to the store for bread. Ford was determined to reintroduce Bronco as a strong, confident, and durable vehicle that could go anywhere, anytime, and would cultivate an avid following of customers who advocated that lifestyle.

The sightings of new vehicles healed old wounds and restored faith for the legions of existing Bronco fanatics, and Ford enthusiasts in general, that the model line was finally, truthfully, coming back. Years of rumors and innuendos were now fact.

Unequivocally the prototypes were met with delight, evidenced by customer response and orders. That so much of the vehicle pays homage to its original ancestor, with its taut bodylines, a familiar front end, and the option of only two doors, was nothing short of a Christmas miracle. What made the worldwide launch even more audacious was that there wasn't just one model, but two: Bronco and Bronco Sport! To add further excitement, the promise of the Bronco Sport Badlands edition suggested that the baby Bronco was going off-road too. It was gratifying to see all the First Editions snapped up as quickly as they were.

FORD'S MARK GRUEBER AND TED RYAN ON BRONCO

Given all this drama, which seemed to approach levels of hysteria, why did it take Ford 24 years to officially announce Bronco's return?

A discussion with Mark Grueber, Bronco Marketing Manager, and Ted Ryan, Ford's Archives Manager, provides some insight.

The Bronco Four-Door Outer Banks Fishing Guide was purpose-built to support a professional fishing guide. This lifestyle concept is based on the regionally inspired four-door Bronco Outer Banks series. On top is a Bestop Sunrider first-row soft top, factory-style concept roof rails and crossbars, and a Yakima LockNLoad Platform roof rack. The SUV makes use of fender-mounted trail sights to fit a custom-made fishing pole and seat perch. A Ford Performance modular front bumper and safari bar help push through rugged terrain, while a slide-out tailgate provides a great work surface. © 2023 Ford Motor Company.

Bronco Sport was offered on Day 1 as an alternative to the larger Bronco. It fits a specific niche, and its corresponding lower price tag has contributed to its overall appeal in addition to its five original trim levels. (First Edition obviously is no longer available, replaced by the Bronco Sport Heritage and Heritage Limited Editions). © 2023 Ford Motor Company.

BRONCO VS. BRONCO SPORT

The Ford Bronco family currently comprises two distinct models. Why? First, there are two distinct price points. Second, and more likely, two distinct target audiences. Third, one vehicle is for multiple purposes and the other a daily driver. Fourth, to be able to sell *as many vehicles as possible*.

And other noticeable differences?

Size?

The advantage goes to Bronco Sport for parking and maneuverability, unless one chooses the two-door bigger Bronco, which is just one inch (3 cm) longer.

Chassis Preference?

Bronco is body-on-frame construction, while Sport is a unibody crossover. Still, Ford wants owners of both to believe their vehicle is capable of on off-road excursion and all-around adventure.

Horsepower?

Bronco is available in two-door and four-door body styles. The SUV comes with two engine options: the 2.3L (140 cu in) I4 EcoBoost with 300 hp (221 kW) with 325 lb-ft (441 N·m) of torque or the 330 hp (243 kW) with 415 lb-ft (563 N·m) of torque. Bronco Raptor is powered exclusively by the 3.0L (183 cu in) V-6 EcoBoost capable of 418 hp (307 kW) and 440 lb-ft (597 N·m) of torque. *Car and Driver* claimed its test of a Raptor made it from 0 to 60 in 5.6 seconds.

Sport utilizes Ford's unibody C2 platform (along with Escape, Focus, and others) and offers two engine options: the 1.5L (92 cu in) I3 EcoBoost and the 2.0L (140 cu in) I4 EcoBoost. Standard with the Base, Big Bend, and Outer Banks trims, the 1.5L (92 cu in) is rated at 181 hp (133 kW) and 190 lb-ft (258 N·m) of torque, while the 2.0L (140 cu in) engine puts out a respectable 250 hp (184 kW) as well as 277 lb-ft (376 N·m) of torque. The latter comes standard exclusively on the range-topping Badlands trim. Tests have indicated this package can scoot from 0 to 60 in about 7.2 seconds.

Gas Mileage?

If a Raptor is in the mix, then fuel economy does not make it a contender, but Sport does boast better numbers than its bigger sibling. The base engine generates an EPA estimate of 26 miles per gallon combined or 11 kilometers per liter (KPL), while the 2.0L 140 cu in) offers 23 MPG (10 KPL). Bronco ranges in the neighborhood of 17 to 20 MPG (7 to 9 KPL), given its engine size and driving conditions.

Carrying Capacity?

Clearly, size matters. Bronco is bigger than Sport. However, Sport has been configured in such a manner than while it may be only an inch (3 cm) smaller than the two-door Bronco, it's absolutely an animal in storage capacity with 32.5 cubic feet (920.3L). Bronco four-door has 38.3 cubic feet (1,084.5L).

Towing?

If one wanted to tow, would a Sport even be in the equation? Bronco is essentially a pickup, so that works, but Sport is a crossover and towing isn't its strong suit at just 2,200 pounds (998 kg). Bronco can pull 3,500 pounds (1588 kg), which is not a tremendous amount, but tow capacity is neither's strength.

Interior Luxury?

Price factors in here. Sport is on dealer floors to arguably appeal to first-time buyers or for buyers of their first new vehicle. It doesn't offer a tremendous number of interior options or big tech. Nevertheless, being a crossover, Sport does have a quiet, comfortable ride and a lot of standard optional features that will play on its standard eight-inch (20 cm) touchscreen.

Bronco, being a truck at heart, rides a little rougher, and the bigger the tire, the noisier the ride may become. The interior simply by the size of the rig is going to be larger, and the further up the trim levels one goes, the higher the quality in terms of materials, finish, and offerings.

Adventurousness?

Seriously, both win here. Whether or not the owner wants to go off-roading, both Bronco and Sport can generally get there and back, but Bronco will go further by capability. It was built for that purpose, and again, the higher the trim level, the more it can climb, crawl, and leave pavement behind. But Sport is a capable little dynamo in its own right.

Ford focused on Bronco Sport Badlands as the off-road-oriented Bronco Sport, powered by the 2.0L (122 cu in) four-cylinder EcoBoost, mated to the eight-speed auto. It embraces all the goodness that overlanding offers and has towing capacity of 2,200 pounds (998 kg). © *2023 Ford Motor Company.*

Interestingly Grueber started with Ford in June 1996, the same time the fifth-gen Bronco was discontinued. "However," he stated, "there were several attempts by internal Ford teams—three major ones, as I recall—to resurrect the Bronco, but each did not succeed. It was the fault of the economy and overall feasibility.

"Demand for new vehicles is customer- and market-driven," Grueber noted. "What you also need, of course, is a plant with open capacity, a platform, and a strong business case with leadership support who will prioritize it. The attention for many years was not on Bronco and we didn't have a place to build it anyway.

"However, things change, and the car market was changing," Grueber continued. "We knew we were going to bring back the Ranger to North America. And when the Ford Focus was pulled out of the Michigan Assembly Plant, we had capacity. The market was there.

"The last year of Bronco sales—despite going away—seemed like it was becoming more popular," noted Ted Ryan. "In the time it was gone, an entire generation grew up without one, but the passion for Bronco remained."

"We're selling more Broncos now than 'in the day,'" Grueber stated. "Clearly, the market over time moved from body-on-frame [construction] to [unibody] crossovers. I think we've made a case for a rugged vehicle that 100% of our audience today prefers. What's more, the original Bronco didn't have a lot of options, and we looked to Ford and our experience in motorsports to be able to offer customers the ability to make modifications. It was definitely not the case to take the older Broncos off-road among most of our buyers.

"Now, people want to get into the outdoors," Grueber emphasized. "Research shows overlanding is an area of growth, and this definitely applies to the sixth generation of Bronco. We definitely had that customer in mind when we launched the vehicle."

Grueber went on, "For example, Everglades is the type of vehicle to go deep in the wild. Each model has a purpose. Bronco historically has been a two-door, but we shifted to a four-door to accommodate new customers, yet it was important to still offer that two-door, even if sales of the four have been the majority. Our goal is to make Bronco for the individual, to satisfy all customer needs and uses, from bare bones, where you can wash out the floor, to a loaded one—Outer Banks, First Edition—with every creature comfort.

"It's critical to have the breadth and accessories to allow people to customize," Grueber maintained. "And we want to build upon the powerful brands Ford has created. We have one with [F150] Raptor, and it was natural to extend that to both Bronco and Ranger. Our plan is to keep Bronco fresh and continually offer new derivatives. We'll move forward to offer a high-speed off-roader, too." (That's most likely the new $300,000 Bronco DR [Desert Racer], announced in mid-February 2023.)

"I think it's fair to say that we've managed to maintain the DNA of Bronco," Ryan added. "It has always been one of Ford's leading brand nameplates. We've never forgotten that. Just like F-Series has such tremendous value. So does GT, and Bronco always had a place. Ford's goal was to make it a capable off-road vehicle. Durable. We incorporated the same spirit as the first generation. Our designers went back to the first strategy papers; dug up the first TV ads . . . the goal was to fully understand what we were dealing with, and then examine all five generations and glean the best of Bronco over all those years.

"In other words," Ryan continued, "Ford's team of designers and engineers first wanted to know 'what made a Bronco a Bronco?' Everyone looked for key design elements, specific characteristics, and then incorporated the original G.O.A.T. modes that encompassed the early development and made Bronco such an icon.

"Today's Bronco can do 100 times what the early one can. And we're proud of that," Ryan declared.

A late trim arrival was the Everglades, which was considered a special-edition model, meaning its package was defined by the vehicle. Though it was based on the Black Diamond trim, this Eruption Green Bronco ultimately bundled unique items to make it a separate offering. © 2023 Ford Motor Company.

Race-prepped for two years, the nearly $300,00 2023 Bronco DR became available for consumer purchase: a turnkey, brutal, high-speed off-road runner. Only 50 Bronco enthusiasts (or racers) can buy one, and it isn't even street-legal. © 2023 Ford Motor Company.

Grueber was quick to add a plug for Bronco Sport. "[It's a] smaller SUV that's more affordable and still a capable, rugged vehicle. It's designed for an active audience who wants to go outdoors, get to the trailhead, do some hiking, biking. It has its niche for sure."

Both men came across as pleased with the results of where Bronco sits at this point. For Grueber, it has been a long time coming and the satisfaction of seeing Bronco become reality and to be as successful as it is currently is clearly evident. Grueber doesn't appear to be someone who is ever complacent, however, and the wheels inside him are constantly turning, preparing a new round of trim models, thinking already about a mid-gen update, whenever that may be. For now, though, the table has been set and it's time to reap the harvest.

For Ryan, it's all about adding to the legacy and building upon Bronco's tremendous achievements. Long may they continue.

As these design sketches of Bronco Sport indicate, Ford execs knew that customers want to get outdoors. "This was definitely not the case with the first gen among most of our buyers," stated Mark Grueber, Bronco Marketing Manager. "Research shows overlanding is an area of growth, and this definitely applies to the sixth gen of Bronco. When it comes to Bronco Sport, the smaller, affordable SUV is a capable, rugged vehicle. As Grueber notes, "It's designed for an active audience who wants to get into the outdoors." © 2023 Ford Motor Company.

Ford never lost sight of bringing Bronco back, always searching for the right time, space, and design. And when all three came together, Ford did it right, with a trio of offerings—Bronco two-door, four-door, and Bronco Sport—making it a complete family.

Chapter 7

Bronco Gets a Return ENGAGEMENT

When it was announced for the first time in 2004 that Bronco "could" come back, well, the reaction was seismic. It registered everything from giddiness to dubiousness. The latter won.

But that didn't mean Ford wasn't thinking about a possible return. In 2004, it wasn't inevitable, but there was a persistent determination to see this vehicle get a reprieve. It would take time and another 10 years and one more prototype to get the design and concept flowing through the corporate veins.

By 2017, it was official.

"When we set out to develop the new car, we knew it had to look like a Bronco and it had to work like a Bronco," said Paul Wraith, chief designer for Bronco and Bronco Sport.

It was clear from the outset that the new Bronco would bear a strong resemblance to the original Bronco, so much so that it bordered on reverence. Round headlights were an absolute necessity, and the shape of the vertical rear end echoed its 1960s roots. And then there was the color-contrasted roof. Bronco was now categorized a mid-size SUV, no longer a full-size as its predecessors from 1977 to 1996.

With hundreds of thousands of requests for information and a corresponding number of orders, or at least reservations placed, Ford knew it had its work cut out for the factory folks. Union relations are always on the border of contentiousness, but Ford has enjoyed a good partnership with the United Automobile Workers (UAW), and now the manufacturer was going to call upon its incredibly skilled personnel to assemble this symbol of renewal whose pioneering forebears had established a path for all future off-roaders. It was a cherished moment when the factory fired up with the new tooling. Bronco first gen had come down this same line and now 56 years later, the sixth gen would become a reality.

And then, there's a pandemic.

By 2021 the connection with the past was very evident, even highlighted, as Ford insisted not only were its new stable of Broncos "Built Wild," but they maintained a strong tie to its previous kin. © 2023 Ford Motor Company.

CHOICES AND CHOICES AND CHOICES

To be fair, this was June 2021, and while the pandemic hit much earlier in March 2020, it continued to impact daily life and in Bronco's case, production. Shortages affected all assembly lines, and consumers learned that a dreaded little semiconductor was slowing down their dream. Broncos and their Ford brethren were stacking up outside factories waiting for the valued part in order to complete their journeys to their respective Ford store throughout the country.

Flash forward to 2023 and customers are still waiting for their Broncos. Ford has tried to mollify each and every one by notifying them of their vehicle's production date, sending them an image of its completion, and updating them on shipping and pending delivery. Most buyers are happy, judging from their postings on various Bronco sites, and are delighted with their Bronco upon its arrival.

One aspect of the entire launch of Bronco sixth gen that was unprecedented was the fact that the vehicle was offered in so many different configurations, also known as "trim levels." Each version really made Bronco a unique personality.

Underneath, of course, away from the eye, what designers call "under the skin," came the updated componentry that made Bronco sixth gen run right: body-on-frame construction with front twin A-arm independent suspension and at the opposite end, a rear five-link coilover suspension and solid Dana axle. Customers could check the option box for a HOSS—a high-performance off-road stability suspension system that included "position-sensitive" Bilstein shocks with multiple compression and rebound zones at all corners. A highly modifiable front sway bar was installed that could be hydraulically disconnected to increase articulation should the driver take the Bronco rock crawling, and it would automatically reconnect when the vehicle went back to normal speed conditions.

Of five packages offered, Standard, Mid, High, Lux, and Sasquatch, it was the Sasquatch platform that became increasingly popular. It included a suspension lift, 17-inch (43 cm) black-painted aluminum wheels and 35-inch (89 cm) mud-terrain tires, electronic-locking front and rear axles, the above-referenced Bilsteins, 4.7 final-drive ratio, and high-clearance suspension complemented by similar fender flares, all perfectly engineered for off-roading. Bronco

Those who looked closely may have noticed the series of positions the bucking bronco logo went through during its existence as the brand's mascot. Call it a refinement or a tuning, but whatever the basis behind it, the horse did change its stance over the course of progression.

It started as a script logo and then graduated to the horse known and loved by customers worldwide. The bucking bronco was designed in 1964, and judging from this Ford-archived image of a stylist diligently carving out the original horse, the similar looking Mustang logo was used as either a guidepost or a deterrent.

These images clearly show the style of the era in which they were applied to their respective Bronco. The newest logo is a slight upgrade from that which appeared on the 2004 prototype at that year's Detroit Auto Show (as indicated on its accompanying brochure).

What remains clear is that all Ford personnel involved in the rebirth of Bronco paid homage to its heritage and ancestry, down to the footprints it left.

A stylist provides final details to the clay mock-up of what will become Bronco's enduring horse. This was completed as early as 1964. © 2023 Ford Motor Company

Given out as the press kit for the 2004 Bronco prototype was this superb mock-up brochure showing a slightly modified bronco standing up straighter in the front. The logo again reflects the age, with rounded letters more condensed and bolder. *Courtesy of the author.*

Bronco sixth gen was built right featuring body-on-frame construction with front twin A-arm independent suspension and at the opposite end, a five-link coilover suspension and solid Dana axle. Customers could check the option box for a HOSS, a high-performance off-road stability suspension that included "position-sensitive" Bilstein shocks with multiple compression and rebound zones at all corners. A highly modifiable front sway bar was installed that could be hydraulically disconnected to increase articulation for those planning a day off-roading. The sway bar would automatically reconnect when it was time to return the Bronco to regular street duties. © 2023 Ford Motor Company.

designers played up the significance of the G.O.A.T. modes, which marketing called its Terrain Management System and developed as optional equipment, in order to maximize traction when buyers chose to go on the trail. Seven modes—besides normal (daily driving)—can be utilized through this system including Eco (good for you), Sport (getting more active), Slippery (be careful), Sand/Snow (watch your paint in the sand, clean it off if you're in the snow), Mud/Ruts (live for the dirt), Baja (now we're talking) and for pros only, Rock Crawl. What else could one want?

Well, there are two more driving options. The Trail One-Pedal Drive automatically applies and holds the brakes when the driver lifts off the gas pedal, thus removing the need for left-foot braking and most importantly, preventing unintended rollbacks, an option well worth the price when climbing crests that aren't normally attempted. The second is a Trail Turn Assist that utilizes the vehicle's new torque-vectoring system, helping Bronco turn in tight corners when off-road. Everyone needs a little assistance to overcome the challenges of dirt, rock, snow, and sand.

Inside the cabin the driver/enthusiast can toggle between 4WD Low, 4WD High, 4WD Automatic, and 2WD High (rear-wheel drive), controlled by a dial near the gear selector rather than a secondary shifter that was the traditional lever.

This 2021 Iconic Silver Outer Banks four-door Bronco interior is highlighted by Ford's terrain management system, a.k.a. GOAT (Goes Over All Terrain), a declaration whose roots go back to before Bronco first gen was ever officially approved. GOAT offers up to seven modes in order to maximize traction when enthusiasts want to take their Bronco off-roading. *Photo by James Lipman/ jameslipman.com.*

Designers had a heyday when they conceptualized the Badlands with the Sasquatch package, playing off the aggressive but mystical nature of the concept. Accordingly the Sasquatch package quickly became one of the most popular Bronco options and could be ordered on nearly every model. © 2023 Ford Motor Company.

This 2022 Shadow Black Badlands two-door Bronco, with black-painted modular hardtop constructed as an homage to the original Roadster, features removable doors and frameless windows allowing Bronco sixth gen to live a life of open-air enjoyment. © *2023 Ford Motor Company.*

So much of the original Bronco flowed through the sixth generation. In an homage to the Roadster, removable doors with frameless glass combined with a detachable roof allows Bronco sixth gen, as well as its occupants, to live a life of open-air movement without inhibition, although properly secured within their two- or four-door model. Especially significant is that Ford stylists wanted the new Bronco to be completely user friendly for all genders, and the doors can be easily removed and stored in the rear cargo area. From a safety practicality, the side mirrors remain attached to the vehicle.

Certainly no journey can be without additional digital technology, and Ford has a plan for that too, offering up a software package that provides a thorough topographical map called "Trail Maps," which allows travelers to download maps and directions to Bronco's in-car entertainment system as well as record videos of trail runs, display telemetry, and map data to its monitor while also uploading the collected information to the cloud. For the true purist, a mount is available on the front dash for phones and mobile cameras, and waterproof switches and rubber floors are available. If the mud and dirt come in, there's always a way to wash it out. Those front fender raised brackets, which Ford calls trail sights and says they help with navigation, are more likely intended to serve as tie-downs and further accessory mounts.

Bronco sixth gen knows no boundaries, which was actually the advertising tag line ("No Boundaries") used by Ford for its SUVs in late 1999. Ford's current tagline is "Built Ford Proud."

(continued on page 88)

FAITH-BASED AND CHARITABLE BRONCOS

In 1980, as noted in Chapter 4, Ford modified a Bronco for Pope John Paul II during his visit to the United States in October 1979. The *Popemobile* had a unique look and safely transported the Pope throughout the country.

History repeats itself, and in 2021, Ford and its Ford Motor Company Fund committed to building another special Bronco that they would donate for charity, in this case with all proceeds to support the Pope Francis Center in Detroit, Michigan. This particular

Bronco First Edition—a rare vehicle in its own right—followed many of the styling cues as its 40-year-old predecessor, fielding Wimbledon White paint with matching steel wheels and one-of-a-kind Rapid Red striping on the doors and decals on the fenders symbolizing it was the "First Edition Pope Francis Center."

This Bronco also carried a Ford Performance lightbar on the roof, sidepod lights, Rigid wheel-well rock lights, and tube doors. Interior seats

wore white vinyl and bore *Bronco* lettering in red. An in-vehicle safe with a MOLLE strap system mounted to the inner swing-gate complemented the trim.

"It's an honor to be able to help support a great charity like the Pope Francis Center," said Steve Gilmore, Design Chief, Ford Vehicle Personalization, in preparing the vehicle for its moment in the sun.

The center itself is a homeless facility, offering food and shelter as

The new *Popemobile* is actually a Ford-built and -donated new Bronco sold at auction with proceeds donated to the Pope Francis Center in Detroit, Michigan. © 2023 Ford Motor Company.

well as hygiene services and medical and legal clinics to those in need. Its name was chosen to honor the Catholic Church's present pope but has served Detroit for more than three decades.

Known internally to Ford as *Bronco 66*, the vehicle—originally donated by David Fischer Jr., President and CEO of the Suburban Collection Holdings, LLC, a major dealer group—was sent upon its completion to the huge Barrett-Jackson auction in Scottsdale, Arizona. On January 26, 2022, the regal Bronco (Lot #3001) sold for $500,000, all of which went to benefit the Pope Francis Center. Talk about a heavenly intervention.

Bronco Wild Fund

As Bronco was just getting off the ground in late 2020, Ford went on the marketing and charitable offensive by developing the Bronco Wild Fund, a new endowment funded by a portion of proceeds from Bronco sales that intends to support the "responsible enjoyment and preservation of the great American outdoors." The mission focuses on "a deep respect for public lands" and will concentrate on reforestation, trail management, outdoor adventure, and scholarship programs.

The goal of the campaign is to raise up to $5 million each year. In its first year, Bronco Wild Fund's ambition was to plant one million trees. Bronco Brand Marketing Manager Mark Grueber said part of the fund's

purpose is to enable Bronco owners and off-road enthusiasts in general to become "responsible stewards of our nation's treasures."

An ongoing policy among off-roaders over the years has been to protect the environment while enjoying nature at its best. The "tread lightly" motto sprung from this devotion. Bronco Wild Fund intends to work via strategic alliances with nonprofits selected for their preservation efforts of America's outdoors, including the National Forest Foundation, which is deeply invested in reforestation, and Outward Bound USA, which enables young people to experience the outdoors through its specially organized programs.

Since its start in 2020, Bronco Wild Fund has since added partners America's State Parks, Sons of Smokey Wilderness Defense, and treadlightly. org, and has committed grants and money to each.

The National Forest Foundation launched the 50 Million for Our Forests campaign to raise funds by 2023 in an effort to plant 50 million trees across the National Forest System, and Bronco Wild Fund continues to support that goal.

Ford also follows its own path toward stronger sustainability (see the "There's Green Inside the Blue Oval" sidebar) and uses many recycled materials within vehicle production in addition to moving deeper into electric vehicle (EV) manufacturing.

Bronco will likely play an even greater role in the overall strategy once it too joins the EV lineup.

The special Ford-created badging—designed to mimic similar Bronco identity on the fenders—is a one-off for this special vehicle designed for a unique mission. © 2023 Ford Motor Company.

(continued from page 85)

TRIM LEVELS

At its original announcement, the 2021 Bronco was available in six options that were broken down into the following:

Base

The entry-level trim of the Bronco is known as the Base (some customers, in keeping with the whole "Built Wild" theme, call their model "Basecamp" and have even gone so far as to produce some "suggested" images to identify their vehicles as such. The standard rendering identifying this model, or trim level, is simply the horse.

When first announced, the manufacturer's suggested retail price (MSRP) was pegged at $29,995. For 2023, the starting price bumped to $34,595.

Available as either a two-door or a four-door model, the 2022 Base came with the 2.3L (140 cu in) EcoBoost twin-scroll turbocharged inline four-cylinder engine (300 hp [221 kW] with 325 foot-pound [lb-ft] or 441 Newton meters [N·m] of torque) mated to a seven-speed manual transmission. That's six gears, plus one "crawler gear."

From a handling standpoint, standard is the four-wheel discs and ABS, plus the HOSS 1.0 system. Options are the HOSS 2.0 system with the Bilsteins (described previously) and high clearance ride height and other upgrades such as trail one-pedal drive, which requires the 2.7L V-6 (165 cu in) and Trail Control, necessitating the 10-speed automatic. Standard wheels are the 16-inch (41 cm) bright polished silver-painted steel and P255/70R16 30-inch (76 cm) tires. Upgrades are endless and all get bigger, glossier, and obviously, pricier.

In the cabin, one gets cloth seating and manual six-way positioning for both driver and passenger. A/C is standard as is an AM/FM stereo with seven speakers and subwoofer, Ford's SYNC 4, featuring an eight-inch (20 cm) touchscreen with wireless Apple CarPlay and Android Auto compatibility, SiriusXM Satellite Radio, AppLink, 911 Assist, and FordPass Connect.

As can be expected in today's mobile society, multiple charging ports are available, and many of the expectations of a modern vehicle are incorporated: keyless entry, push-button start, precollision assist with automatic emergency braking, manual tilt and telescoping steering column, and various everyday conveniences, as well as carpeted flooring, sliding visor vanity mirrors, second row window controls on the two-door, and *Bronco* badging throughout. Plus, one

can't miss Ford's very clever Terrain Management System offering five G.O.A.T. modes.

Upgrades begin with floor mats or floor liners, cargo area protectors or cargo area rug, slide-out tailgate (on the four-door), cargo area storage, safe deposit/console lock box, and tailgate table.

There's more standard stuff: manual swing-out gate, molded-in-color door handles (or black), black molded-in-color grille with *BRONCO* lettering, heated sideview mirrors, and tow hooks. Options include a Carbonized Gray molded-in-color hardtop for the four-door as well as retractable full top or front row, both in twill soft material; a canvas Bimini top; soft tonneau cover (4-door); top mesh Bimini shade; two- or four-tube doors; brush guard; heavy-duty modular front bumper; body appearance kit; and body armor protective moldings.

From a safety perspective, Ford offers its internally designed Personal Safety System, which provides an overall level of frontal crash protection to front seat occupants by assessing the size of the occupant in the passenger seat as well as crash severity to determine what airbags to deploy during a collision. Multiple positioned airbags are included along with a Belt-Minder system in the front row. Other standard features include a perimeter alarm, LED headlights, AdvanceTrac Roll Stability Control (RSC), and for those considering towing, a trailer sway control. Only a keyless-entry pad and prewired auxiliary switches in the overhead console are options from this aspect.

Finally the 2.7L (165 cu in) twin-turbo V-6 begs to be added, offering 330 hp (243 kW) and 415 lb-ft (563 N·m) of torque, and the 10-speed automatic transmission and more advanced four-wheel-drive system come along with it. You can also affix the automatic transmission to the four cylinder.

And, like icing on a cake, Ford allows customers to tie its Sasquatch Package on even the Base model, and that continues up through the trim levels.

For 2023 key highlights keep the focus on Bronco's 4×4 with part-time selectable engagement—electronic-shift-on-the-fly (ESOF). The two-speed transfer case is the base part-time system that delivers three modes for specific driving conditions plus neutral. Each selectable mode increases wheel torque to handle conditions like deep sand, steep grades, or heavy pulling.

The class-exclusive cowl-mounted side-view mirrors stay in place when doors are removed. This of course is offered throughout the line. The hardtop retains its ability to be easily removed and is standard on two-door, while a folding soft-top goes standard on the four-door.

Big Bend

The Big Bend is named as a tribute to the famed national park in Texas, close to Carroll Shelby's ranch. It is, like all sixth-gen Broncos, available as either a two- or a four-door model and adds more convenience and styling features to the Base trim, including 17-inch (43 cm) aluminum-alloy wheels finished in Carbonized Gray with 32-inch (81 cm) tires as well as a leather-wrapped steering wheel and gear lever. Clearly, as one moves up the price range, the vehicle becomes more upscale. Initially priced at $34,880 for the two door and $37,375 for the four, the two-door now lists for $38,585.

The base engine remains the four-cylinder EcoBoost. The 2.7L (165 cu in) V-6 can be added as well as the Sasquatch Package. Much of the options available to the Base can be ordered for Big Bend, though there are more, such as heated seats and a voice-activated touchscreen navigation system as well as the Mid Package, which is a number of options bundled together, including many of the features previously listed as interior options on Base, plus Ambient Footwell lighting, Dual-Zone Electronic Automatic Temperature Control (DEATC) reverse sensing system (a.k.a. rear parking sensors), connected navigation, a 110-volt power outlet wireless phone connection, cloud-connectivity, Adaptive Dash Card technology, personal profiles, *digital* owner's manual, lane-keeping alert, and driver alert. Ford's Co-Pilot360 suite of driver-assistance systems also is included.

Overall it makes for a significant upgrade and much more comfortable and accommodating vehicle as well as a safer ride.

Black Diamond

The term *Black Diamond* can have two meanings. In the jewelry business, black diamonds are incredibly unique and represent inner strength and justice. Some jewelers fashion them for clients who want to stand apart or stand strong in their convictions. But like any diamond, they symbolize a strong emotional connection. And from a skier's perspective, the black-diamond run is the steepest available, the toughest possible, a narrow slope with more hazards and high winds throughout its trail.

In naming this third level of the Bronco multitiered system, Ford marketing and design personnel wanted to convey that Black Diamond is both rugged as well as rare while appealing to individuals who want an exceptional vehicle.

Big Bend editions are next on the option ladder.
© 2023 Ford Motor Company.

The Black Diamond trim package is level number three.
© 2023 Ford Motor Company.

However it was identified, the "Diamond" was definitely not "in the rough," even though it offered the driver the chance to play in the same, especially with seven G.O.A.T. modes this time round. Ford called it "next-level outdoor adventure" with features such as "marine-grade vinyl-trimmed seating services" that worked well in this scheme, plus rock rails, heavy-duty skid plates, and a powder-coated steel front bumper with LED front fog lamps and tow hooks to light up the way and pull to one's heart's content, respectively. A companion steel rear bumper is installed too. A clever interior highlight is the rubberized flooring with drain plugs.

Also on deck are 17-inch (43 cm) black-painted steel wheels with 32-inch (81 cm) tires as standard with the Advanced 4×4 with automatic-on-demand engagement and the Sasquatch and Mid packages available as options. Black Diamond two-doors are listing at $41,250 in 2023.

Outer Banks

This package is named for the incredible barrier islands off the North Carolina shoreline where Orville and Wilbur Wright got their start. The brothers conducted their first flights off Kill Devil Hills, located on Bodie Island, which provided them with a wide expanse of open beachfront, frequent winds, and an agreeable climate.

The Outer Banks trim level therefore is a mix of stylishness and practicality, with 18-inch (46 cm), gloss black-painted aluminum alloy wheels mated to 32-inch (81 cm) tires, signature LED headlights and taillights, fog lamps, body-color painted fender flares, and power-coated tube steps. Ford wanted to outfit its Outer Banks Bronco with both off-road capability and improved road manners and fit them in a more luxury-oriented package that included dual-heated front bucket seats and the Mid Package as standard. There were six G.O.A.T. modes. Still optional was the 2.7L (165 cu in), 10-speed auto, Sasquatch Package, along with new choices befitting this level, such as leather-trimmed seats and the High and Lux Packages. Initial prices were set for $38,995 for the two-door and $41,450 for the four. In 2023, Outer Banks starts at $44,155 for the two-door.

Outer Banks—named for the island community that includes Kill Devil Hills, where Orville and Wilbur Wright first earned their wings—is the fourth option. © 2023 Ford Motor Company.

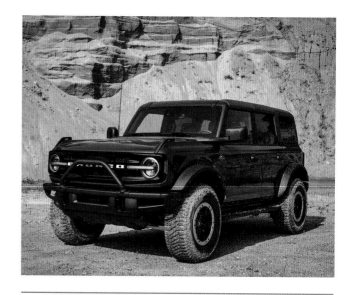

Eruption Green has a strong appeal, particularly when displayed on the four-door Outer Banks model. © 2023 Ford Motor Company.

Badlands

Named after the Badlands National Park in South Dakota (or a Bruce Springsteen song), Bronco's Badlands version presents one of three off-road-oriented Bronco trim levels. As with its siblings, Badlands is available as either a two- or four-door model and expands upon features found on the Outer Banks model such as 17-inch (43 cm) aluminum-alloy wheels finished in Carbonized Gray with 33-inch (84 cm) tall tires, additional G.O.A.T. modes for the four-wheel drive system, an upgraded suspension system, a front stabilizer bar disconnect, and the powder-coated steel front bumper with integrated LED front fog lamps and tow hooks, and it includes the marine-grade vinyl-trimmed seating surfaces. The 2023 list price sits at $47,395.

Wildtrak

The Wildtrak is another off-road-oriented Bronco trim level. Available as either a two-door or a four-door model, the Wildtrak trim adds features onto the Badlands trim level such as the Sasquatch Package, the 2.7L (165 cu in) EcoBoost twin-turbocharged V-6 gasoline engine mated to a ten-speed automatic transmission, a unique front hood graphic, cloth seating surfaces, and dual heated front bucket seats. Initially set at $43,590 for the two-door, the 2023 model is now running at $55,830.

(continued on page 95)

Wildtrak is another off-road-oriented trim level. © 2023 Ford Motor Company.

Badlands conjures the tough but amazing National Park in South Dakota and implies this version is ready to take on just about any terrain. © 2023 Ford Motor Company.

Four-door Wildtrak with optional HOSS 3.0 that includes the Fox Internal Bypass Dampers Package. © 2023 Ford Motor Company.

BRONCO EASTER EGGS

What exactly is an Easter egg besides the traditional one you boil and paint? According to a variety of sources, it's a term referring to any number of curious items hidden or unexpected, usually presented as a treasure, surprise, or treat.

Ford's designers put an amazing amount of work into the new Bronco. Aside from the assortment of trim options and accessories available from the Blue Oval, there were a number of subtle applications that became a part of the Bronco identity and legend.

These are the design Easter eggs that adorn hidden spots throughout several trim levels of Bronco.

Bottle openers appear on both Bronco sixth gen and Bronco Sport. These are useful little items. The other Easter eggs are primarily imagery, although with very clever messaging tied to it. For example, on the roll cage when the tops are removed is a mountain range stamped onto a piece of black trim with coordinates 34.52621 N and 116.75685 W. A quick search via findlatitudeandlongitude.com tells the adventurer that this point is near Twentynine Palms, Joshua Tree, and Johnson Valley OHV Area, the location of the annual King of the Hammers race, where the new Bronco was first unveiled in racing mode (see Chapter 13).

Many publications, both print and online, as well as owners, have discussed the number of hidden Easter eggs that Bronco holds, but Wendel Motors, a family-owned and -operated dealership in Spokane, Washington, serving Ford customers since 1943, put together an extensive list of 51 items including the aforementioned bottle opener and "Bronco Knoll" location in Johnson Valley.

With the dealership's permission, an edited version of the list is included here:

Horse in the Headlights

1. The Ford Bronco is "Built Wild" using the famous bucking bronco as its logo that's reverentially etched into the edge of the LED daytime running headlights.
2. The front Bronco letters are lit with amber backlighting.

Trail Sights

3. *Bronco* is engraved in the bolts.

Hidden Louvres

4. On the Bronco Raptor, the hidden louvres on the hood have years imprinted on them—1967, 1969, 1971, and 1972—indicating the years that the Bronco won the Baja 1000 race.

Front Bumper

5. Each of the bolts has *Bronco* engraved on them.

Wheel Well

6. *LIFT ME BABY* is embossed in the front wheel wells.

Sasquatch Package

7. The Wildtrak Sasquatch trim badge has a Sasquatch walking through.
8. *Wildtrak* is written at the bottom of the vehicle.

Accessory Ready

9. The Bronco comes ready to add lots of accessories with reference to this "opportunity" noted throughout the vehicle in the form of "Accessory Ready" labels in specific areas (see Chapter 11).
10. All the accessory bolts are labeled with *BRONCO* and *MNP 10.9*.

Optional Slide-Out Rear Cargo Panel

11. The panel has an engraved first-gen Bronco with a stick figure sitting on the tailgate enjoying a campfire.

Rear

12. The bucking bronco is on the rear end of the SUV.

Taillights

13. The lights are shaped like horseshoes, and the passenger side taillight bears a strong resemblance to the letter *B*.

Tires

14. While this may not be an actual visual Easter egg, it's something that's "missing" from the Goodyear tires. The tires on the Bronco are Goodyear's Wrangler line of tire. For what might be obvious reasons, Ford preferred not to carry the "Wrangler" name on exterior side of the rubber. Instead, the name *GOODYEAR* appears twice on the outer sidewall, with Wrangler appearing on the inner sidewall. Take that, you guys with seven slats in your grille!

Wheel Center Caps

15. Each center cap has everyone's favorite bucking bronco.

Windshield

16. *BRONCO EST. 1966* is on the windshield, making sure everyone knows that the original Ford Bronco is the inspiration for the new Bronco.
17. The QR code is the production date, used for the manufacturing process.

Gas Cap Door

18. Open the door to view three 1966 Broncos with the U-codes of the first-generation Bronco models: U13 for the roadster, U14 for the half-cab pickup, and U15 for the wagon. In the Ford Bronco Raptor, the graphics are of historical race-winning Broncos that competed in the Baja 1000 race.

Door Connectors

19. The door wiring connection cover has *GET UNHINGED* engraved on it.

Tailgate

20. Open the pull-out tray in the cargo area to see the Ford tailgate script with the same font and pattern of the original 1966 Bronco.
21. The tailgate glass has a Bronco with a surfboard hanging out the back end.
22. There is only one Ford emblem on the vehicle, located on the bottom of the tailgate on the driver's side.
23. On the Raptor model, open the tailgate and mounted on the inside of the third brake light is the word *RAPTOR*, printed backwards. Drivers can read it in their rearview mirror.

Rear Wiper

24. The latch panel for the glass that holds the rear wiper motor assembly has a mountain range etched on it.

Rear Bumper

25. Behind the rear bumper license plate is *BRONCO EST. 1966*.

Grab Handle

26. The interior grab handle has *BRONCO* embossed on it.

Interior Hardware

27. The bolts holding some of the dashboard and other interior items are aluminum Torx bolts that have *BRONCO, MNP 8.8* engraved on them.

Start/Stop Button

28. The shape of the button is identical to the Bronco's headlights.

Drive Modes

29. For the original Bronco project as discussed in Chapter 1, Ford used an internal code name of G.O.A.T., which referenced "Goes Over Any Terrain." On the new Bronco, G.O.A.T. is obviously used for the driving modes.

Inclinometer

30. The inclinometer will show you the incline of where you are driving using a first-gen Bronco. You can see the terrain, backwards, forward, and all around.

Interior Lights

31. The light in the cargo area has an illuminated bucking bronco on it.

Tie-Down Points

32. The western heritage of the Bronco wild horse is celebrated with a lasso symbol engraved on the four tie-down points in the rear cargo area.

Hooks

33. The hooks in the cargo bay have lassos embossed on them.

Badge

34. Beneath the shifter is a badge that states that the vehicle was designed and engineered in Dearborn, Michigan.

Center Console

35. *No Step* is etched in the console.

Driver's Screen

36. When the screen displays a starry night, stars will shoot across the screen.

Center Screen

37. As the screen shows boulders rolling, they will then take the shape of the bucking bronco.

Cubbyhole

38. Pull out the rubber pad in the cubbyhole under the radio. Turn it over and depending on the model, a scene of a mountain range and trees or a mountain range with cactus will be displayed.

Air Conditioner Vents

39. The A/C vents are shaped like the letter *B*.

Silhouettes

All the silhouettes in the vehicle are of the Bronco, rather than a generic vehicle.
40. Center screen
41. Traction control off button
42. Air recirculation button
43. G.O.A.T. mode switch
44. Hood release switch
45. Disclaimer sticker inside the door

Shifter

46. The shifter has the Bronco logo and an American flag on it. The vehicle is proudly built in the United States, after all.

Glove Box

47. Inside the glove box is the packet of tools used to take off the doors with a Bronco logo on them.

Rear Seats

48. Open up the bottom of the seat and there'll be a hidden compartment with a shoe and first aid kit stamped inside.

Second-Row Seats

49. The armored rubber backers that snap on the rear seats are engraved with *BRONCO EST. 1966*.

Trunk Area

50. As noted, a bottle opener is conveniently located on the rear portion of the roll bar.

Roll Bar

50. When the tops are removed, a mountain range is engraved on the trim stating the coordinates 34.5261 N, 116.75685 W. These points, which call out "Bronco Knoll," lead to Johnson Valley, California, home of the King of the Hammers race, where Ford sponsors races and has done extensive testing of Bronco. On some models, these are located inside the lift gate on the side.

(continued from page 91)

First Edition

The 2021 First Edition model was perhaps the cleverest concept of the group. This was—at the time—the biggest and most exclusive package, which made it incredibly desirable because it was a controlled production, and in today's society, any item, especially vehicles, produced in a limited supply, becomes immediately collectible.

Based on the mechanical equipment of the Badlands, Outer Banks interior, and Wildtrak exterior and available in two- or four-door, the First Edition included most options as standard equipment, including Mid, High, Lux, and Sasquatch Packages, and was initially limited to just 3,500 units, although Ford later doubled the number of these special Broncos due to increased demand. With special graphics, including a First Edition hood and Shadow Black-painted hardtop, Safari bar, and leather seats (and power driver seat), there wasn't an option box left. Final production run remained 7,000 units, and First Edition was not offered again. It wasn't cheap, either, at $60,800 for a two-door and $64,995 for the four-door. Imagine what these are selling for now on the aftermarket (though surprisingly, buyers seem to be holding onto theirs).

2021 Bronco First Edition interior. Many special cues indicated the First Edition was unique to the lucky few who purchased one. Within the interior were such symbols as Bronco signage on the dashboard, colored door accents, and First Edition badging on the console. It was clearly one-of-a-kind. © 2023 Ford Motor Company.

Everglades

A late arrival to the number of offerings for 2022 was the Everglades trim named for the 1.5 million acres (6,070 km^2) of wetland in South Florida. Announced in December 2021 for availability in Summer 2022, Everglades Bronco was available as only a four-door, unlike its siblings, and identified as a special-edition model. It was based on the standard equipment found within the Black Diamond trim, plus unique front fender graphics that doubled as a depth chart should the vehicle be used in water. Also included within the package were special aluminum-alloy wheels, heavy-duty modular front bumper, and upgraded electronics. Everglades are starting at $54,645.

Ford Bronco Everglades in Desert Sand. Bronco Everglades feature special Ford-installed accessories to help increase its water fording capabilities. © 2023 Ford Motor Company.

Raptor

Finally, as a sort of New Year's 2022 gift, in another attempt to capture press and public attention along with new customers, Ford announced the one Bronco upgrade pretty much everyone had been waiting for, the Raptor version. Quickly nicknamed, or renamed, *Braptor*, the new four-wheel drive beast on the block easily transcended what most buyers wanted. Available as a four-door only, Raptor became the top-tier Bronco.

Set for Summer 2022 launch, the Raptor was built to appeal to high-performance fanatics as well as steal more market share from the manufacturer up the road in Auburn Hills (think seven-bar grille). At this point, the Raptor line was made up of pickups: the F-150 and more recently, the Ranger, both of which were designed to integrate off-road trophy truck components onto street-legal variants. The success Ford has encountered from its F-150 Raptor alone justified the investment in a Ranger version and obviously secured placement within Bronco. Raptors are starting at $76,580.

ABOVE: Tough and terrifying—to those who either don't have one yet or to the competition. Raptor just brews confidence throughout. © 2023 Ford Motor Company.

BELOW: There are four models of Heritage Edition Broncos: Bronco Heritage Edition and Heritage Limited Edition and Bronco Sport Heritage and Sport Heritage Limited Edition, offered in both two-door and four-door versions. © 2023 Ford Motor Company.

Heritage Editions

Last but certainly by no means least is the latest announcement and newest sibling. The Bronco and Bronco Sport Heritage Edition and Heritage Limited Edition models were announced in August 2022 as 2023 releases.

Like First Edition and Everglades, these vehicles have been designated as special editions but will be offered as either two- or four-door versions. Ford will build just 1,966 units—in honor of its very first year of production of the original Bronco—of each Heritage Limited Edition model in both Bronco and Bronco Sport.

Indeed the premise behind the Heritage Edition is to commemorate the first generation, "paying respect to Bronco's roots with nods to some of the signature design cues from that vehicle," stated Mark Grueber, Bronco's Marketing Manager. He also hinted that Ford would continue to expand the Bronco brand with more special editions that customers want.

Accordingly Bronco Heritage models are highlighted by a two-tone paint job that includes signature Oxford White accents and similarly colored grille with Race Red

FORD lettering. Seventeen-inch (43 cm) aluminum heritage wheels also painted in Oxford White and a bodyside stripe dial up the throwback looks. According to Ford media releases, "Heritage Limited models add gloss black-painted 17-inch (43 cm) heritage wheels with classic "dog dish" centers, plus an Oxford White–painted lip for more retro flare."

Inside Bronco Heritage Edition units feature plaid cloth seats and unique touches such as an Oxford White dashboard, center console badging, and exclusive front and rear floor liners. Big Bend trim is standard along with the 2.3L (140 cu in) EcoBoost engine with either seven-speed manual or 10-speed auto. The Sasquatch Package is also included.

Bronco Heritage Edition is available in five paint options, while Bronco Heritage Limited Edition is exclusively available at launch in Robin's Egg Blue, which is a throwback color based on Arcadian Blue, available on the original Bronco in 1966. The color Yellowstone Metallic, which is based on the 1971 Ford color Prairie Yellow, is planned for late 2023 model year availability. Peak Blue is planned for the 2024 model year.

(continued on page 100)

COLORS, OH MY!

You want proof the designers, stylists, marketers, and execs at Ford aren't having a wee bit of fun with Bronco sixth gen? How about color choices? Well, first, check this survey.

According to the study "America's Most Popular Car Colors12" by iSee-Cars.com, the top colors are (drum roll unnecessary because the answers are predictable and thus, boring): white (28.5%); black (22.3%); gray (18.4%); silver (12.1%); blue (9.5%); and red (8.6% and mainly Ferraris and Corvettes). All vehicles painted in other colors are under one percent.

This is where one says, "C'mon, people, be imaginative!" However, it isn't the consumer's fault. Colors are driven by the manufacturer. There are some automakers who will offer you a custom choice and for a hefty fee (hello, Porsche), provide a palette of colors from which the buyer can select.

But Ford seems to have gotten it right with Bronco. And the names are as cool as the colors they represent.

Not all models could get every swath, but in 2021, the choices were:
- Rapid Red Metallic Tinted Clearcoat
- Velocity Blue
- Shadow Black
- Antimatter Blue
- Iconic Silver
- Area 51
- Carbonized Gray
- Cactus Gray
- Race Red
- Cyber Orange Metallic Tri-Coat (late arrival)
- Oxford White

Exclusive to the First Edition models was a vibrant Lightning Blue. It's the way things are done when they're done right.

Now the author (and probably most people) has never been to Area 51, so there is some curiosity as to how Ford colorists know it's some shade of dark gray. And how does Antimatter Blue have any color at all if it's, in fact, antimatter?

Just thinking out loud, of course.

New colors for 2022 were:
- Eruption Green
- Hot Pepper Red Metallic Tinted Clearcoat

Those returning were:
- Area 51
- Shadow Black
- Cactus Gray
- Carbonized Gray
- Iconic Silver
- Oxford White
- Race Red
- Velocity Blue

Gone was Antimatter Blue (it did disappear, apparently) and Rapid Red.

New for 2023 were:
- Azure Gray, which got a resounding, "Really, another gray?" But it has a blue tint to it, so that apparently makes a big difference.
- Antimatter Blue came back. By request?
- Area 51 did not return for 2023. Wiped clean, no doubt.

That's it. Returning for 2023 were:
- Eruption Green
- Hot Pepper Red Metallic Tinted Clearcoat
- Shadow Black
- Cactus Gray
- Carbonized Gray
- Iconic Silver (dead last among color choices, according to Bronco6gen)
- Oxford White
- Race Red
- Velocity Blue
- Area 51

Other colors that were discontinued:
- Cyber Orange Metallic
- Code Orange—This is available only on Bronco Raptor, as well as the Mustang Shelby GT500 and F-150 Raptor. As they should.

For 2023 Bronco Sport, there was some good news:
- Area 51 is still available as well as Cyber Orange Metallic. Finally, something just for Sport.

Additionally, the following could be requested in 2023:
- Eruption Green
- Atlas Blue Metallic
- Hot Pepper Red
- Shadow Black
- Iconic Silver
- Alto Blue
- Carbonized Gray
- Cactus Gray
- Oxford White

Whatever one's capacity for uniqueness, Ford has come a long way since founder Henry used his great marketing aptitude in saying "any customer can have a car painted any color he wants so long as it's black."

LEFT: The 2021 Badlands with Sasquatch package in Cactus Gray. © 2023 Ford Motor Company.

OPPOSITE: Raptor in Hot Pepper Red Metallic Tinted Clearcoat. *Courtesy of the author.*

The newest sibling within the extended Bronco family is the Heritage Edition and the even more exclusive Limited Edition. Just 1,966 units of the latter will be built (along with a Bronco Sport version). Naturally each version will receive special trim packages and distinctive identification. © 2023 Ford Motor Company.

Heritage Editions are as important to Sport as they are to the rest of the Bronco line-up. All Sport models feature signature Oxford White accents including a painted roof to mimic the original Bronco style. An Oxford White heritage grille features Race Red lettering, separating Sport from the bigger Bronco. © 2023 Ford Motor Company.

(continued from page 97)

Heritage Limited Edition models offer the increased capability of Badlands series equipment and features, including a 2.7L (165 cu in) EcoBoost V-6 engine with up to 330 horsepower (246 kW) and 415 lb-ft (563 N·m) torque mated to the 10-speed SelectShift automatic transmission. They are also differentiated with features that include metal *Bronco* script fender badging, leather-trimmed/vinyl plaid seats with white and Race Red accent stitching, and unique *Heritage Limited* console badging.

BRONCO SPORT PROUDLY WEARS THE BRAND LEGACY TOO

All Sport models feature signature Oxford White accents including a uniquely painted roof to mimic the original Bronco style. An Oxford White heritage grille features Race Red *BRONCO* lettering, separating Sport from the bigger Bronco. Sport also wears Oxford White 17-inch (43 cm) aluminum heritage wheels. Additional accessories are dictated by whether the vehicle is a Heritage or Heritage Limited Edition featuring considerable upgrades but with callbacks to previous generations of Broncos.

Bronco Sport Heritage Edition is available in seven paint options including Robin's Egg Blue, while Heritage Limited Edition is exclusively available in Robin's Egg Blue, Yellowstone Metallic, and Peak Blue.

Bronco Sport Heritage Edition is listed at $34,245 while Sport Heritage Limited starts at $44,655. Bronco Heritage has an initial price tag at $45,555 and Heritage Limited begins at $68,145.

For the non-Heritage Bronco Sports, there were five trims initially available: Base, Big Bend, Outer Banks, Badlands, and First Edition.

Base came with the 1.5L (92 cu in) three-cylinder EcoBoost engine connected to a rotary-controlled eight-speed automatic transmission. Big Bend became the mid-level model of Bronco Sport, adding mostly convenience upgrades, while Outer Banks was identified as the luxury-oriented trim level, including 18-inch (46 cm) tires, aluminum-alloy wheels, combination leather-and-cloth-trimmed seating, and a Shadow Black-painted front grille with white *BRONCO* lettering.

A 2021 Cyber Orange Metallic Tri-Coat Bronco Sport First Edition properly identified through its badging. Every Bronco carries telltale signage indicating what trim level it represents and image it symbolizes. © 2023 Ford Motor Company.

Moving to Badlands, Ford focused on this model as the off-road-oriented Bronco Sport, powered by the 2.0L (122 cu in) four-cylinder EcoBoost, still mated to the eight-speed auto. Two additional G.O.A.T. Modes were added to Badlands Sport four-wheel drive system, which also included all-terrain tires and unique aluminum-alloy wheels.

Finally Bronco Sport received its own First Edition, which was only available for the 2021 model year, with components based off the Badlands trim as well as virtually all of Sport's available options and packages, including leather-trimmed and dual-heated front bucket seats and upgraded sound system. The First Edition trim package was limited to just 2,000 units, making it the rarer of the two First Editions.

Ford also offered bundles where it packaged trim levels on Sport with specific accessories, calling each lifestyle accessory bundle a separate name: Bike, Camping, Snow, Water, and Cargo. Clearly, something for everyone.

BRONCO SCOUT?

FordAuthority.com, an extensive source for Ford and related brand news, reviews, research, and other information that's led by Alex Luft, reported on the filing of trademark applications at the U.S. Patent and Trademark Office and discovered on April 15, 2019, before Bronco had begun production, Ford had indicated its interest in registering the name "Scout." The application was listed for "land motor vehicles, namely passenger automobiles, pick-up trucks, sport utility vehicles."

Of course, the name Scout was owned by Navistar, the successor company to International Harvester, but Volkswagen bought the former in July 2021 and in May 2022 announced it was reviving Scout for a new electric off-road vehicle. Former Audi of America President Scott Keogh is running the operation now.

And what became of Ford's application for a trademark for Scout? It was abandoned by Ford Attorney Sharon C. Sorkin on September 9, 2019.

Prior to Bronco arriving at MAP, Ford invested $750 million in 2021 and added 2,700 more jobs to prepare the factory for the new SUV. Here a Bronco comes down the line followed by a Ranger, painted in the same Cyber Orange Metallic Tri-Coat, both just about ready for testing.
© 2023 Ford Motor Company.

Chapter 8

Building the NEW BRONCO

I t's been over 100 years since Henry Ford built the innovative River Rouge Plant, called "The Rouge," in Dearborn, Michigan, and since then, Ford has maintained a loyal commitment to all its facilities—and to its employees.

Wayne, Michigan is host to Ford's 2.8 million square foot (260,128.5 square meters) Michigan Assembly Plant (MAP), formerly known as Michigan Truck Plant, and currently home to approximately 4,900 hourly employees and 300 salaried workers. The city has seen millions of brand-new vehicles roll through its community since 1957 when the factory opened.

For 65 years, many of the local citizens have more than likely worked at the 369-acre (1.5 km²) plant that originally opened as the Michigan Station Wagon Plant building the Mercury Colony Park station wagon.

A formidable platform in its day, station wagons remained the only production duty until the plant was retooled to produce pickup trucks in 1964 and became the Michigan Truck Plant, beginning with the F-100. A year later, 100,000 trucks had claimed residency due to the skills of the labor force within the factory. Profits from the sales of those vehicles led to further investments in the manufacturing facility in 1968, increasing capacity. Further expansions occurred in 1974, 1991, and 1996, each time adding capacity as well as new models to the production line.

Bronco came aboard in 1966 and remained through its original five generations, resulting in a name revision to the factory as the Michigan Assembly Plant. At the end of that 31-year run, Ford and its factory workers had produced more than 1.1 million Broncos.

GREEN INSIDE THE BLUE OVAL

Ford has made it a corporate mission to become more sustainable and environmentally friendly throughout its organization. This goal rang true particularly among the manufacturer's factories. In October 2012, *Assembly* magazine awarded the Michigan Assembly Plant (MAP) as the first factory in Michigan and the first auto plant to earn Plant of the Year honors for its flexible, green manufacturing and transformation capabilities in order to efficiently move from producing large SUVs to fuel-efficient small cars.

Indeed Ford's commitment to eco-friendliness has become a core value as reflected in the manufacturer's placement of grass on the roof of its historic Rouge factory. This 10-acre (4,047 m²) "living" roof is the largest in the world and represents Ford's determined efforts to embody the ecological footprint of a large manufacturing facility. The green cover that rests atop the truck factory contains sedum that insulates the building, lowers energy costs, cleans the air by trapping carbon dioxide, and extends the useful life of the roof.

According to the website Greenroofs.com, "redevelopment of the 1917 complex formed the foundation for the company's vision of balancing lean manufacturing with environmental sensitivity."[4]

Ford Executive Chairman Bill Ford, whose great-grandfather Henry Ford constructed the complex, led with this quote in 2000:

"This is not environmental philanthropy; it is sound business, which for the first time, balances the business needs of auto manufacturing with ecological and social concerns in the redesign of a brownfield site.

"This is what I think sustainability is about, and this new facility lays the groundwork for a model of 21st century sustainable manufacturing at the Rouge," Bill Ford continued. "While most companies would rather move than invest in an 83-year-old site, we view this as an important reinvestment in our employees, our hometown and an American icon of the 20th century."[5]

Ten years postinstallation, North America's largest living roof—about the size of eight football fields—continues to flourish atop Dearborn Truck Plant's final assembly building, part of the Ford Rouge Center in Dearborn, Michigan. Various plants, insects, and animals have come to depend on what is now a thriving ecosystem that equates to a 10.4-acre (42,087 m²) garden. The living roof also serves as a cost-effective alternative to roof maintenance for Ford, while keeping heating and cooling costs in check for the factory beneath. © 2023 Ford Motor Company.

Ford recognized the demand for pickup trucks, and eventually the Michigan Assembly Plant became the site of choice for the F-100. It was renamed the Michigan Truck Plant. This 1964 F-100 is a product of that plant. © 2023 Ford Motor Company.

Ford's Michigan Assembly Plant (MAP) has had a long and varied life, producing cars before trucks. It became the Michigan Truck Plant in the 1960s before being renamed MAP as its manufacturing duties expanded to include other vehicles. *Courtesy of Getty Images.*

The Michigan Assembly Plant (MAP) in 2015 with a Focus displayed outside. This plant has been busy all its existence, taking on new projects, new vehicles—through modifications and retooling to the factory—and delivering quality Ford cars and trucks to millions of customers, thanks to the incredible combined efforts of the line workers and management. © 2023 Ford Motor Company.

MAP had no downtime when Bronco production came to a rest in 1996 as the Expedition immediately took over the assembly line, quickly followed by the Lincoln Navigator and shortly thereafter the Ford Focus and C-Max. SUVs became the focal point until 2010 when Ford completed a $550 million renovation that enabled the plant to quickly modify production runs between multiple models, easily moving from gas-powered platforms to electric to hybrid and plug-in hybrid, without significant downtime—a major financial savings.

By mid-year 2021, as Bronco production ramped up, Ford opened the Modification Center, a 1.7 million square foot (250,838.2 square meter) facility in what was formerly the Wayne Assembly Plant, situated adjacent to the Michigan Assembly Plant (MAP). It wasn't the first "Mod Center," but this one was dedicated to Bronco, enabling customers to—at their discretion—request additional upgrades or options that can be installed either once their Bronco comes off the assembly line or subsequent to delivery.

Within this complex comes the availability of roof racks, front bumper safari bars, exterior graphic packages, different tire sizes (for appropriate applications), and lights. Most of the items are on display, and the new Bronco owner can select what they want and have those items installed on site.

It's a tribute to the modular design of the Bronco that allows for as many accessories to be fitted easily—Ford has created more than 200 factory-backed products thus far—giving buyers the ability to personalize their vehicles either through their dealers or independent sources.

In fact, the aftermarket has responded in kind (see Chapter 12) with a huge inventory of parts, trim pieces, and functional equipment that add more power, prestige, and versatility to every Bronco and pride to the owner. All of it comes at a cost, but it's whatever demand will allow. When a product is hot, the entrepreneurs and creatives arise, introducing both practical and ornamental items that are all designed to make one's Bronco original, even more than when it emerged from the factory. Whether it's an easy grille change, adding larger fender flares, or making major modifications, Bronco is positioned as an adaptable machine ready to be reconfigured by customer choice.

Custom Broncos. A major part of the excitement at the annual Easter Jeep Safari in Moab, Utah, is viewing the many prototypes manufacturers introduce to off-road enthusiasts. Ford uses the event to reach out to its target customers—who are there to drive various trails—but also to see what's new from the Blue Oval as well as new custom builds by Ford and its aftermarket suppliers. © 2023 Ford Motor Company.

RANGER BACK HOME

In 2017 Ford announced its mid-size Ranger would return to North America and be produced at MAP. The truck was offered with this nameplate in other parts of the world, built by Ford Australia, but remained unavailable to buyers in the United States and Canada since its cancellation in 2011. This was great news for the factory workers and the city of Wayne. The factory would get a new life with an old brand plate—actually, two. Bronco was coming along too.

On October 22, 2018, the celebration of a rebirth began. The occasion marked the return of a pickup to the former Michigan Truck Assembly factory, but more importantly, it signaled the arrival of a thoroughly modernized mid-size truck that American consumers demanded. After investing $850 million to retool the plant and taking just four weeks to install new equipment to accommodate both Ranger and Bronco, Ford executives pushed their chips to the center of the table. The iconic vehicles would increase Ford's truck and SUV leadership in North America and augment two of the company's profit centers. The company also invested $150 million at the Romeo Engine Plant nearby to expand its capacity for engine components.

Building the Ranger began for the 2019 model year and Bronco arrived in 2021, filling the factory quickly, though production of Bronco was delayed until the end of summer 2021 due to the dual difficulties of COVID-19 and supplier issues.

Nevertheless production of the all-new 2021 Bronco began on June 15, 2021, following another $750 million investment by the Blue Oval and the addition of 2,700 direct jobs at MAP.

By that point, more than 125,000 orders had been placed for Broncos with a total of 190,000 reservations across North America. "Broncomania" was in full swing.

At launch date, both management and labor were on the same side of the table. "We have the most skilled workforce in America, working in a plant that's state-of-the-art," said John Savona, who was at the time Ford's Vice President, Manufacturing and Labor Affairs.

Savona was supported by Gerald Kariem, former UAW Vice President and Director of the Ford Department (now retired and replaced by Vice President Chuck Browning), who added, "Once again, UAW members are excited to build one of the most iconic vehicles in automotive history as the full-size Ford Bronco launches production."

MAP started building Rangers in 2018 after the factory received an $850 million upgrade and retooling.
© 2023 Ford Motor Company.

This beautiful Azure Gray Metallic four-door Bronco is fired up and ready for its postproduction test drive.
© 2023 Ford Motor Company.

As fast as Ford is building its internal combustion engine (ICE) Broncos, the corporation is equally accelerating its internal transition toward electric vehicles, and no doubt—though officials are not giving time frames—Bronco will eventually bear its own battery bank at some juncture down the road.

The new plan is called Ford+. Not very original, much like the streaming services, but this is a very detailed and creative strategy. It divides the company into distinct operations: the Ford Model e division dedicated to electric and connected vehicles, billing Ford "as the center of innovation and growth," and the Ford Blue division, which remains focused on producing "iconic Ford vehicles and experiences," while serving as the "engine that supports and powers Ford's future."

Of particular note, Ford Model e will take the lead on creating an exciting new shopping, buying, and ownership experience for its future electric vehicle customers that includes simple, intuitive e-commerce platforms, transparent pricing, and personalized customer support from Ford ambassadors.

And over at Ford Blue, this little nugget appeared within its announcement: "Ford Blue will exercise Ford's deep automotive expertise to strengthen the iconic Ford vehicles customers love, such as F-Series, Ranger and Maverick trucks, Bronco and Explorer SUVs, and Mustang, with investments in new models, derivatives, experiences, and services."

It's pretty safe to say that everyone wants to know just exactly what these "derivatives" might entail.

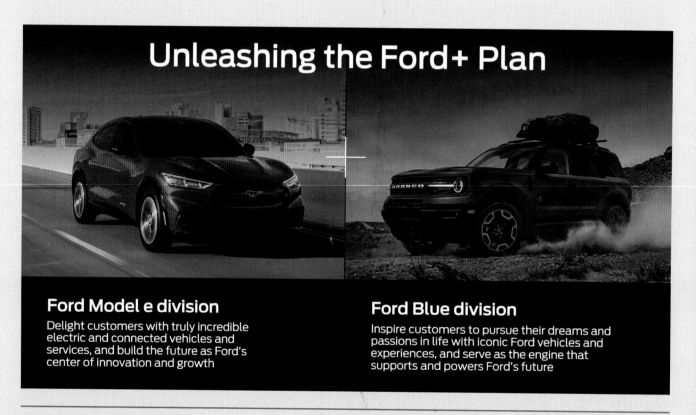

Unleashing the Ford+ Plan

Ford Model e division
Delight customers with truly incredible electric and connected vehicles and services, and build the future as Ford's center of innovation and growth

Ford Blue division
Inspire customers to pursue their dreams and passions in life with iconic Ford vehicles and experiences, and serve as the engine that supports and powers Ford's future

Ford continues to transform its global automotive business, accelerating the development and scaling of breakthrough electric, connected vehicles, while leveraging its iconic nameplates to strengthen operating performance and take full advantage of engineering and industrial capabilities. © 2023 Ford Motor Company.

EXTERNAL CHALLENGES

Unfortunately the plans didn't progress as rapidly as forecast. The pandemic saw to that. This first started within the plant but also with supplier Webasto, which provided various roof components. The company had coronavirus-related delays starting in December 2020 that resulted in late deliveries to Ford, which had to shift the start of production to June 2021. Once the Bronco roofs arrived for assembly, quality issues with the hardtop hampered production and began the setbacks that put the entire manufacturing process behind. Ford pushed for quick resolution and by early December 2021, had replaced all hardtop roofs at MAP and was working diligently with existing Bronco customers and their dealers to install a hardtop roof replacement.

But as most consumers are aware, the real nail that put a flat in the assembly line's movement was the global shortage of semiconductors and related supply chain challenges, a major plague that continues to impact Ford and other manufacturers' factories. As a result of minimal supplies of these crucial components, Broncos and sister vehicles built at eight Ford plants were forced onto the sidelines, unable to be shipped as they awaited the critical part.

> *"The global semiconductor shortage continues to present challenges to a number of industries— including automakers worldwide. Our teams continue prioritizing key vehicle lines for production, making the most of our available semiconductor allocation, and will continue finding unique solutions around the world so we can provide as many high-demand vehicles as possible to our customers and dealers. Ford will build and hold these vehicles for a number of weeks, then ship the vehicles to dealers once the modules are available and comprehensive quality checks are complete." Statement from Ford on the chip shortage—May 4, 2021[3]*

The semiconductor shortage became an extensive story in both 2021 and 2022. While certainly not as significant as COVID-19, its complications unleashed their own kind of harm. It resulted in a domino effect within the automotive industry, pushing production to its limits, forcing cutbacks and delays, and triggering tough political conversations between countries regarding responsibility for maintaining production of the critical chips in the first place.

Nevertheless, Ford and its dealer body continued to receive tremendous demand for Bronco. In June 2022, order banks for all Blue Oval vehicles swelled to 300,000, resulting in the closure of many new orders across a number of lines. Ford made it clear that it also would cancel any Bronco reservation that had not been converted to an order by November 21, 2022. Matt Winter, Bronco Brand Manager, spoke at length in late January 2023 with the leaders of Bronco Nation, which telecast the discussion live on its YouTube channel, about the Vehicle Order Holder Information and what would next occur for the 2022 model year and beyond.

MOVING FORWARD

At that time, there were 37,000 outstanding Bronco orders, and Winter stated it was the factory's goal by early 2023 to clear that backlog. He added that shipping times had improved and he expressed confidence that MAP personnel were working as quickly and as hard as possible to build everyone's order.

It's true that many Bronco enthusiasts waited a long time, going on more than two years, for their Bronco to arrive, posting on social media when they receive an email from Ford indicating their order is in the "queue," when they receive a VIN (vehicle identification number), and also when Ford has delayed their delivery date.

However, Ford has been good about sending images of customers' vehicles when they're completed and ready to ship. Again this is a way of building a closer relationship with them, and Winter added during the interview how much he has "appreciated everyone in the Bronco community."

"We're really glad to have people who want to elevate Bronco to a special level," he noted, recognizing the importance of getting their vehicles to them sooner rather than later. He stated that the reservation process by 2024 would likely mature and change, so Ford can "give everyone who has been waiting for a Bronco the opportunity to get one, as we provide our customers the best equipment and the best Broncos they can get."

---

[4] Ford Motor Company press release, 2000.

[5] http://www.greenroofs.com/projects/ ford-motor-companys-river-rouge-truck-plant/

Bronco II production began in January 1983 and was ready for sale at Ford dealerships by March of that year. It was identified as a 1984 model—though some have argued that early production numbers were referenced as 1983½. It's possible that these may have been preproduction test vehicles. © 2023 Ford Motor Company.

Chapter 9

A Tale of
TWO CHASSIS
Bronco II and
Bronco Sport

When a brand is successful, its handlers or managers—the marketing folks—want to take advantage of that recognition and loyalty. In the case of packaged goods, that means creating additional shelf space options, and for the auto industry, offering more choices in the showroom. According to any marketing textbook, this is commonly known as a *brand extension*. As long as the brand's original name is included in the extension, it generates brand recognition, trust, and ultimately, incremental sales.

With the main Bronco selling well since its redesign in 1978, Ford designers, engineers, marketers, and sales personnel believed that a smaller Bronco, sized closely to the first-generation version, would do well as a second model under that label, just with an added identification of "II." There also was the Mustang II that during its four-year run sold one million cars and was considered "the right car at the right time."

As far as brand extensions were concerned, the consensus might have been, if Coca-Cola can do it with Diet Coke (and later, Vanilla, Cherry, and Zero) along with M&M's with its myriad selection including plain (later renamed milk chocolate), peanut, dark chocolate, almond, crispy, peanut butter, and caramel, for a total of 59 flavors at present, then certainly, a successful automotive manufacturer like Ford could.

BRONCO II (1984–1990)

The Bronco II was introduced in the 1984 model year, though some consider it technically a 1983½ (or 1983.5) due to its spring release. Much like its cousin's first design, the II nicked parts from other models, including the Ranger. The vehicle was built to reach loyal customers, but also to compete against the competition launched the previous year in 1983, including the Jeep Cherokee and the Chevrolet S-10 Blazer and its GMC twin, each offered as compact two- and four-wheel drive three-door wagons. This, of course, was before they were called SUVs.

All Bronco IIs were four-wheel drive until 1986 when rear-wheel drive became standard. Earlier models ran a German-built carbureted 2.8L (171 cu in) Cologne V-6 boasting 115 hp (85 kW) at 4,600 RPM, the same engine placed within the 1984 and 1985 Ranger, plus a four-speed manual transmission. In addition to the new RWD system, the 1986 Bronco II received a fuel-injected 2.9L (177 cu in) Cologne V-6 engine that produced 140 hp (103 kW), a valuable increase in power. Curiously Ford released a Mitsubishi four-cylinder 2.3L (140 cu in) turbodiesel as an option during the 1986–1987 production years, but it was poorly received and dropped.

Bronco II was conceived as another string to Ford's violin. The new SUV was intended to partner the Ranger in the Blue Oval's goal to offer a condensed line of vehicles. Economies of scale were at play here with shared platforms, facades, and engines. Consumers who didn't want the larger Bronco and who dreamed of a smaller Bronco again would be in the market for this type of sport utility. Naturally the new sibling offered more improvements over the original generation Bronco and made for a more appealing choice. And it could tow a trailer weighing up to 4,050 pounds (1,837 kg).

What, exactly, was the Bronco II's "true north?" What was it designed to do?

"When you find your true north, you discover your authentic self. It's a combination of your purpose and your beliefs. Get on the right course, proceed in the right direction." [6]

As evidenced by the sales numbers over Bronco II's six-year life span, its principal purpose was to rope new customers into dealers' showrooms. And that plan succeeded.

The marketing for Bronco II was very reflective of a conservative era, with copy that declared it was a "vehicle for men, single people, or young couples"—a return to the G.O.A.T. concept that was equally comfortable in an urban setting. It was successful with first-time buyers and for those looking to add a second or third vehicle to their garage. People who had bought the original/first gen Bronco were reliving their youth with Bronco II or buying one for their children who were likely driving age by this time or were on their own and remembered the joy of riding in the family Bronco years before. Ford stylists and marketing personnel developed multiple trim levels as well: XL, XLS, XLT, and Eddie Bauer. By September 1984 several running changes

Bronco II had various trim levels. This first-year model was identified as an XLS, which was one step up from the base version and offered additional upgrades, including accent stripes. © 2023 Ford Motor Company.

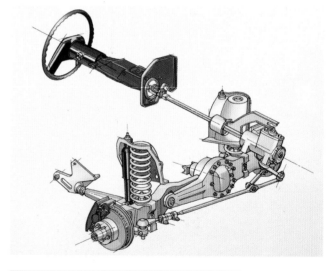

When first introduced in 1984, all Bronco IIs were four-wheel-drive. In 1986 rear-wheel-drive became standard. The first two model years saw vehicles equipped with a German-built carbureted 2.8L (171 cu in) Cologne V-6, the same engine used in the 1984 and 1985 Ranger, backed by a four-speed manual transmission. A Dana 28 front axle was part of the Twin-Traction Beam (TTB) independent front suspension. © 2023 Ford Motor Company.

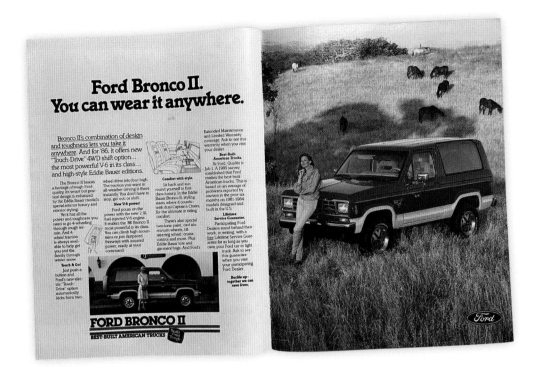

were introduced that gave each Bronco II version a distinctive look, upgrading the models beyond their introduction.

Part of these changes were based on the fact that Bronco II had to satisfy other goals too. Some of them were logical, such as emissions reduction and competition to the increasing numbers of smaller SUVs spilling onto the market, and some of them were emotional, such as increasing the number of Bronco faithful. Though those who favored the bigger SUV were typically not the target to convert, its two doors and smart appearance pulled at those who were just beginning to see both the appeal of a sport utility vehicle and the great outdoors. The fact that Bronco II was small, nimble, and could be both a daily driver and an off-roader was a great entry point.

Bronco II was never intended to be a substitute for the bigger Bronco. It was a complementary ride. What's astounding to many, and no doubt to the managers at Ford at the time, is that it consistently outsold Bronco third gen over its life. Some claim it was the V-6. Others state the vehicle's off-roading capability, and still more cite the many manual transmission choices offered (several from Mazda and Mitsubishi) as the model progressed through the years.

Conceivably the most important role in defining Bronco II's true north was that it provided Ford's next direction in the SUV category. Bronco II brought to market several design innovations or attributes that carried forward: the opening glass window in its rear hatch; optional rear reclining buckets instead of a bench seat; an interior spare vs. the optional swing-out tire carrier; electronic shift-on-the-fly transfer case captured through a button accessed via an overhead console; rear anti-lock brakes (introduced in 1987); and a late sport package that featured a front air dam and side skirts.

In 1989 Bronco II and its brother, Ranger, were restyled with new front bodywork, hood, fenders, and a tighter-fitting front bumper. The dash was redesigned with a new instrument panel. Structural support was improved. Four-wheel drive models had Dana 35 front axles installed, replacing the Dana 28. Cylinder heads were improved as well.

Despite this serious investment, Bronco II wrapped production early, in February 1990. What lay ahead for Ford would be its most successful release, the Explorer, which also was based on the Ranger. It came to market just a month later, in March 1990, and while the four-door was

significantly larger than Bronco II, the two-door Explorer, identified as Explorer Sport, bore a strong resemblance and featured many similarities, though it was still 10 inches (25 cm) longer. Nevertheless the lineage was strong, and by 1993 Explorer was selling over 300,000 units.

The introduction of the smaller, less powerful, and cheaper Bronco II proved to be a big success in terms of sales. Although not as capable or durable as bigger sibling Bronco, Bronco II was good enough for most customers. Right from the start, Bronco II began to outsell regular Bronco, and when production ended in 1990, Ford counted 627,304 units sold. It had been a wise gamble.

ABOVE: Storage capacity was significant, with rear seats that individually folded, allowing for multiple applications for carry-ons or carry-thrus. Compartments in the rear sides accommodated more gear. © 2023 Ford Motor Company.

OPPOSITE: From today's perspective, Bronco II was underpowered. Though it was offered with 2.8L (171 cu in) and 2.9L (177 cu in) V-6 engines, they made only 115 and 140 hp (86 to 103 kW) respectively. Nevertheless the little six-cylinder moved the lighter-weight SUV fairly well. Today's engines are smaller yet produce far more horsepower. Interestingly Ford offered Mitsubishi's 2.3L (140 cu in) diesel as an option, but it was soon withdrawn due to poor performance and slow sales. © 2023 Ford Motor Company.

2021 BRONCO SPORT

When Ford relaunched Bronco in 2020 for introduction in 2021, it was a surprise that there would be two models: a larger SUV and one based on the Escape, a logical companion much like Bronco II had been 40 years earlier.

Ford never strays from a winning strategy, even when there are so many changes in management over the years. Creating a new vehicle is an elaborate and expensive procedure; thus, an automaker will extract cost savings where it can. Like its predecessors, Bronco Sport pulled a majority—close to 80 percent—of its parts from Escape and thus, instantly established a very low break-even point. Was it not a real Bronco as a result? Up until 2020, Apple used Samsung and LG screens exclusively for its iPhones.[7] Does that not make it an Apple? Cadillac, Chevrolet, and GMC all use the same platforms for their big SUVs. It's up to each brand to make their vehicle distinct. Same with Apple. If it says it's an iPhone, it is. Ditto with the Bronco.

The strategy is one followed by consumer packaged goods manufacturers. If they can place more of their similarly listed products on the shelf, chances are that customers—familiar with the brand and therefore both loyal and trusting—will reach for it. If Ford offers its potential buyers two versions of Bronco, at two price points, it

Bronco II was fashioned as a new project years after the original Bronco came to market, but Bronco Sport was designed from the outset as a companion to its bigger Bronco sibling and launched simultaneously. © 2023 Ford Motor Company.

Bronco Sport has many adventures of its own, including its own Bronco Off-Roadeo adventure program. Here a 2022 Sport Outer Banks in Hot Red Pepper stands its ground. © *2023 Ford Motor Company.*

will undoubtedly appeal to two different targets: those who want a bigger, more off-road-worthy vehicle, and those simply in need of a great daily driver.

In this case, part of it is the Bronco image that people are buying into. And for this latest generation, the accompanying lifestyle.

While the original Bronco II may have been more seriously aimed at "men, single people, and young couples," Bronco Sport is likely pointed toward *women and* men, single people, and young couples. What's more, the price difference between Bronco and Bronco Sport will draw many customers toward the latter, and while customers will wait for the arrival of the bigger Bronco because it carries a more

impressive image and history, Bronco Sport will be right behind it, for it possesses a legacy of its own and offers great potential at a decent price with a significant profit margin for both dealers and the Blue Oval.

Bronco Sport is all-wheel drive and offers a lockable, torque-vectoring rear axle as well as steel skid plates and can handle itself off road. The round headlights, the Bronco logo on the grille, racing experience, and multiple aftermarket opportunities all contribute to its legitimacy.

What's more, this new Bronco has four doors, something its predecessor never did. It addresses the demand for more space, accessibility, and accommodation—for both passengers and product and equipment. It has so many practical applications, there's a reason its sales continue to expand. Ford announced it is boosting production of Bronco Sport as of March 3, 2023. Total Ford SUV sales across the board have totaled more than 777,000 in 2022, up 4.6 percent from the previous year, and demand will likely continue in the 2023 model year when the Bronco Sport Heritage editions are released.

"Bronco captivated America in the mid-1960s with its rugged style and uncompromising off-road agility, a legacy that is still prevalent today," said Mark Grueber, Bronco marketing manager. "With these new Heritage and Heritage Limited editions, we're paying respect to Bronco's roots with nods to some of the signature design cues from the first-generation vehicle, while continuing to build the Bronco Brand with more special editions that our customers want.

"Bronco Sport has been a hit with compact SUV customers since it arrived two years ago, but it is just as important that we give our customers special editions similar to what we have done with the two- and four-door Bronco," Grueber continued. "These Heritage Editions are every bit as important to Sport as they are for the rest of the Bronco brand, and we think the iconic themes from the '60s work perfectly on this vehicle."

On the heels of this strong association with the original Bronco came another special off-road package for Bronco Sport, "elevating its adventure game," as Ford marketing personnel noted. The package will increase capability while adding a touch of style through the Bronco Sport Black Diamond Off-Road Package, available on Big Bend and Outer Banks series, which includes four steel bash plates that cover key 4×4 powertrain areas such as front metal skid plate, fuel tank, and canister shield. Special 17-inch (43 cm) Carbonized Gray low-gloss lightweight aluminum wheels wrapped by 225/65R17 all-terrain tires provide increased grip in the dirt. From an appearance standpoint, a matte black hood graphic complete with a Bronco horse coupled with bodyside graphics with *BRONCO* lettering distinguish the complete package.

With the increased focus on off-roading at the Bronco Sport level, the customer-centric Bronco Off-Roadeo adventure access program that emphasizes trail guide instruction has been extended in 2023 to include all Bronco Sport customers.

Ford designers spared no expense in creating a comfortable and attractive cabin for the Bronco Sport while placing a high level of functionality at the driver's fingertips. The 8-inch touchscreen with Apple CarPlay and Android Auto compatibility provided ample capacity for driver interactions, while the fashionable dashboard made for a clean view of all gauges. © *2023 Ford Motor Company.*

The all-new Bronco Sport has an available interior bike rack that lets adventurers transport two 27.5-inch (70 cm) bikes inside the vehicle. © *2023 Ford Motor Company.*

A 2023 Bronco Sport Heritage Limited Edition—a highly desirable vehicle—in Yellowstone Metallic will be available in late 2023. © 2023 Ford Motor Company.

Acknowledging the increased interest in moving from highway to back roads, Bronco's Grueber noted, "For adventurers who want to get more from their Bronco Sport, we're enhancing the ownership experience by offering more trail capability with the new Black Diamond Off-Road Package, plus an included opportunity to learn what their SUV can do at Bronco Off-Roadeo.

"Today nearly 90 percent of Bronco Sport customers who attend Off-Roadeo are likely to go off-roading again, and 97 percent of our customers are more knowledgeable and confident doing so, furthering our goal of getting people into the wild."

What sets the newer Bronco apart from Bronco II is the fact that while the latter was built in Lexington, Kentucky, Bronco Sport, along with its Maverick cousin, is produced across the border in Hermosillo, Mexico, at Ford's Hermosillo Stamping and Assembly Plant, which has been in existence since 1986. Bronco Sport shares the C2 platform with fellow Blue Oval products Ford Escape, on which it's based, as well as Maverick, the fourth-generation Focus, and Lincoln Corsair.

The C2 platform, introduced in 2018, is incredibly modular and can be adapted for a number of vehicles and purposes—accepting a torsion beam or multilink rear suspension and applied to varying wheelbases and track widths, from subcompact to compact. Internationally it also is used for the Kuga (the model name for Escape elsewhere in the world), Evos, Mondeo, and Zephyr (the latter two sold in China).

Bronco Sport also has some pretty competitive powerplants: a 1.5L (92 cu in) Ecoboost I3 turbo, producing 181 hp (133 kW), and a 245 hp (180 kW) 2.0L (122 cu in) Ecoboost I4 turbo, coupled to an eight-speed F35 SelectShift Automatic.

This vehicle is quite capable of getting itself onto freeways and out of tough trail routes. Finally, despite the stability troubles experienced by its predecessor, Bronco Sport has been awarded a "Top Safety Pick" by the Insurance Institute for Highway Safety (IIHS).

Scott Evans of *Motor Trend Car Reviews* wrote: "The big Bronco will get all the attention, but the Bronco Sport will bring home the bacon, the same way a GT500 gets buyers into EcoBoost Mustangs."[8]

And he's absolutely right. There has to be one sales leader, and currently, Sport outsells Bronco, but combined, the two SUVs have added a lot to Ford's bottom line, or at least helped cover losses in some areas. Most importantly Sport has brought a new range to Bronco and new customers to the showroom, doing its part to keeping a winning brand on track.

[6] Credit for the term "Discover your true north" goes to Bill George; elements of the quote listed belong to Shonna Water, PhD, cited from https://www.betterup.com/blog/find-your-true-north.

[7] https://illumaware.com/what-cars-share-the-same-parts/. Retrieved on 3 March 2023.

[8] https://www.motortrend.com/reviews/2021-ford-bronco-sport-more-important-than-bronco/.

Bronco Sport has a towing capacity of 2,000 to 2,200 pounds (907 to 998 kg), depending on the engine, so if it's not carrying bikes, it's pulling them off-road in a go-anywhere, do-anything mindset. © 2023 Ford Motor Company.

Chapter 10

Bronco Has a Strong FOLLOWING

. . . BUT DON'T CALL IT A CULT

Technically a product that qualifies as a cult brand by loyal customers (and fans) is one that can do no wrong. It's granted both a long leash and more than three strikes should it bear those out. For example, Apple is a cult brand. People will wait in line in front of the company's stores to be the first to purchase its latest iPhones or other tech. When Apple got itself into a pickle by admitting that it slowed down older iPhones because of battery issues through a software feature it released, there was an outcry that resulted in a series of consumer fraud lawsuits filed by state attorneys general who settled with the tech company for $113 million, a considerable amount. However, since it was Apple, people forgave the brand pretty quickly. Imagine if the same thing happened to an automotive manufacturer. The uproar would last for years, and any settlement would be much, much higher. Obviously there's a reason that Apple is considered the most valuable company in the world, despite this "toe-stubbing."[9]

The fact is, the Ford Bronco could qualify as a cult brand. Despite some missteps both past and present, the SUV has engendered a strong, dedicated audience. Brands that were launched 57 years ago tend to generate a faithful wake, and when the product has a rebirth, as Bronco has, it makes it that much more desirable. Think of a rock band returning to the stage, a legacy band like the Eagles, for example, after being apart for almost a decade and a half. Their comeback resulted in their 1994 album *Hell Freezes Over* becoming number one. Ironically the album name is in reference to a quote by drummer Don Henley after the band's breakup in 1980. Henley was asked in an interview about when the band would play together again, to which he responded accordingly, hence the album title.[10]

Clearly Hell froze over twice as the announcement Bronco fans had waited 24 years finally came. It's one Bronco clubs, online forums, journalists, websites, owners, Ford dealers, used car dealers, aftermarket manufacturers, restoration shops, auction houses, racers, and a broad swath of consumers felt was worth the wait.

This nicely modified 1986 Bronco features a BDS lift kit and 35-inch (89-cm) tires along with a host of tasteful engine and interior upgrades, earning the seller a well-justified $41,600 figure from a lucky bidder on Bring a Trailer in 2022. This generation of Broncos is starting to see increased activity among enthusiasts, raising their respective prices. *Courtesy of Bring a Trailer and seller Tillmanator.*

But it's important to declare here that the Bronco is actually *not* a cult brand. According to Dave Landwehr of Hebron, Kentucky, one of the founders and administrators of the Bronco 6th-Gen Facebook page, an owner of an '88 Bronco and whose wife drives a '21, "people own different vehicles for different needs. My wife and I love our Broncos; her new Bronco is perfect for her, and its independent front suspension was the right set-up for what she needed." But, he was quick to add, for some people, a Bronco has certain limitations and for those reasons, people need another vehicle, such as to pull a big boat.

"So I wouldn't say all Bronco owners are part of some sort of Bronco cult," Landwehr continued. "We use our vehicles accordingly and know that they are made to do specific things. Nothing wrong with that."

Nevertheless the probability that there are more Blue Oval vehicles residing inside the garages of many Bronco owners is quite good. The author, for example, has a Bronco on order, and also has a 1965 Ranchero and a 1966 Mustang at rest within his garage, and like most people who lust after vintage cars and Broncos specifically, seeks an earlier gen U15 to join the brood.

Bronco may not be a cult, but call it a devotion . . . or a reverence . . . or, you get the idea.

BRONCOS CREATE MEMORABLE MOMENTS

It's an evening—and an event—that many baby boomers and any number of Gen X and Y still can easily recall, particularly California residents. The author was eating at a Hamburger Hamlet (now an In-N-Out Burger) across the street from the University of California Irvine with two friends when the TV screen in the bar interrupted the NBA finals basketball game to broadcast live, in what was one of the first televised police chases, a white 1993 Bronco driven by Al Cowlings who was carrying his former NFL team-mate O. J. Simpson cowering in the cargo area (allegedly, since no one could see him, but this is what the broadcaster was saying at the time).

It was June 17, 1994, approximately 6:15 p.m., and Simpson, charged earlier in the day with two counts of murder, had thus far refused to surrender to police. Cowlings claimed he was transporting the ex-NFL star who also was holding a gun (not on Cowlings) and then proceeded to lead multiple police units on a slow-speed, two-hour-long drive up I-5 as well as several other Los Angeles freeways in what was quickly identified as a "police pursuit." The coverage quickly expanded nationally because of the celebrity association and the spectacle taking place (cars stopped along the shoulders and the center of the freeways with people waving and even taking the time to make signs!). To say the author and his friends were riveted to the television coverage, as was everyone else in the restaurant, is like noting that swimming with sharks might prove to be unpleasant.

It was a terrible moment for Simpson, and to this day, a polarizing page in the history of Bronco. The vehicle became instantly recognizable and remains so to this day. Any white fifth-gen (or late-model) Bronco is called an OJ Bronco, and the original Cowlings' SUV ironically sits today as a display piece at the Alcatraz East Crime Museum in Pigeon Forge, Tennessee. It shares space between Ted Bundy's Volkswagen Bug and John Dillinger's getaway car. Considering that Simpson was found not guilty in a criminal court, the white 1993 Bronco shouldn't have any wrongdoing associated with it, so why is it in Tennessee?

And speaking of memorable moments, there's the story of Tracy Conn, General Manager and Partner of Joe Cotton Ford in Carol Stream, Illinois, who is celebrating his 42nd year with the dealership. He started working at the store in 1981 washing cars, graduated to sales two years later, and continued moving up the ladder. Dealer Principal Joe Cotton offered him a partnership in 1989, and he's served in his current role ever since while the store remains family-owned from when it was started in 1974. The dealership's website has an incredibly detailed history of the Bronco from the first generation forward, including covers of each year's Bronco sales brochures through 1977, while Sales Associate Jason Conn conducts a very lively video walk-through of the new sixth generation. It's clearly a dealership that prides itself on building strong connections with its customers and its products and that's important in building a bond with a manufacturer, its offerings, and its buyers.

Many other Ford dealers have done the same thing, with histories displayed on their respective websites along with excellent imagery, some of it provided by the manufacturer, but many photos are original to the dealership, which takes time and effort on each owner's part. Passion plays a significant role, and that's evident in how Bronco is displayed virtually everywhere. A moment of reflection is important in order to recognize the incredible contributions to the automotive industry, and to Ford in particular, by Bert Boeckmann, owner of Galpin Ford, the top volume Ford dealer for 29 straight years, who passed away in May 2023 at 92. He was a tremendous pioneer at all levels within the car business.

And let's not forget how Ford has pushed this new Bronco through its marketing as a "lifestyle" vehicle. The automaker also wisely announced a number of Team Bronco Brand Ambassadors, among them third-generation off-roader Shelby Hall, who graciously wrote this book's foreword (see sidebar Women Who Make Their Mark on the Marque). Shelby, granddaughter of the legendary Rod Hall, has been dedicated to, and consistent in, pushing the Bronco and Blue Oval message for some time now. She also is an advocate of getting more women involved in off-roading in particular by participating in the Rebelle Rally, which runs for eight days every October, and is now in its eighth year.

What's the purpose of this off-road event? Blending one's love of driving with the ultimate challenge of precise navigation, the Rebelle (pronounced Re-Belle) Rally tests entrants' skills over those eight days of competition. It's not a race for speed, but a unique and demanding event based on the combined efforts of proper headings, hidden checkpoints, and time and distance using maps, a compass, and an old school roadbook.

The Rebelle Rally is an all-women competition that covers more than 1,200 miles (1,931 km), beginning near Lake Tahoe, Nevada, and ending at Imperial Sand Dunes in California. It's the ideal terrain to prove the capability and durability of Bronco Sport, which has claimed top honors in its class three consecutive times. © 2023 Ford Motor Company.

Bronco Ambassador Shelby Hall, granddaughter of off-road legend Rod Hall, is a skilled race driver in her own right, competing in various events in both Broncos and Bronco Sports. She's scored victories for the Blue Oval as well as customers and fans through her efforts on behalf of the brand. *Courtesy of Shelby Hall.*

For Ford, competition is often a centerpoint to its marketing ventures, whether car, truck, or SUV. The Bronco was no different and its continuing participation in the Rebelle Rally demonstrates both its strength off-road as well as its popularity among racers and buyers. © 2023 Ford Motor Company.

What's the down-to-Earth focal point of this event? It's the first women's-only rally in the United States. And it's not just an average mainstream rally. It's 2,000 kilometers (that's 1,242.74 miles, to be exact) with no GPS and no cell phone. (More on this and other motorsports events in Chapter 13.)

BRONCO OFF-ROADEO

Ford also conducted an all-woman Off-Roadeo driving event in July 2022 as part of the manufacturer's long-term goal to engage more female customers by showing them the appeal to off-roading as well as communicating more directly with them one-on-one.

The Ford Bronco Off-Roadeo is an annual off-roading program—a one-day "school," if you will—that allows owners of a new Bronco, at no cost to them, get a feel for the SUV's performance on an "outdoor adventure playground."

High-end vehicle manufacturers occasionally treat their customers to driving events, but for Ford to offer this day-long experience to specific verified Bronco customers is proof to the length the Blue Oval will go to cement its connection to its buyer.

Off-Roadeo is hosted in four cities in the United States: Moab, Utah; Gilford, New Hampshire; Horseshoe Bay, Texas; and Las Vegas, Nevada. The women's event was held at Grey Wolf Ranch outside Austin, Texas. Trained Bronco Trail Guides—all female—taught the 27 attendees how to take full advantage of their Bronco's off-road capabilities.

"We wanted to create a space just for [women], so they could comfortably learn about Bronco and ask questions without hesitation," stated Kelsey Gerken, Bronco Consumer Strategy Manager. "The more that we can focus on ownership experiences and pinpoint these different audience to welcome [women] to the Bronco Family and build their confidence level with the vehicles, the better we are able to drive increased loyalty and excitement around not only the Bronco nameplate but Ford in general," she added.

Current owners and order-holders of a qualifying 2021 or newer Bronco; 2021 Bronco Sport First Edition; 2021 or 2022 Bronco Sport Badlands; or any 2023 Bronco Sport can attend before or after taking delivery. A special Bronco Sport program has been instituted with the 2023 model.

Led by industry experts, the unique blend of hands-on and immersive experiences highlights the key aspects that make Bronco an engineering marvel off-road. Participants receive extensive time behind the wheel, driving purpose-built courses that will push their Bronco to demonstrate its engineering technology, especially its G.O.A.T. modes. Attendees also learn a great deal about the rich history of both Bronco and Ford as well as how embrace the outdoor lifestyle and develop their own off-road adventures. It's a brilliant move to further build the community.

Three other Off-Roadeo events were held for both men and women in 2022 with another round of four scheduled in 2023. These programs are free to every Bronco buyer and generally last about 10 hours in a combination of education

Ford enjoyed such success with its Bronco Off-Roadeo programs that it created an exclusive women-only event. Based on the turnout, expectations are high it will encourage return engagements. © 2023 Ford Motor Company

Badges from Bronco Off-Roadeo events of which four locations have thus far been established. Ford-appointed driving professionals appear at these events to assist participants, all of whom are Bronco owners or soon-to-be owners. Participants test-drive a Bronco within their personal comfort zone. © 2023 Ford Motor Company.

(continued on page 130)

The Woman, Bronco, and the Arctic Circle

Once upon a time, there was a woman named Courtney Barber. She lived in South Carolina and was not your average person who goes to school, gets a job, and does the normal consumer lifestyle. Instead Courtney—a Ford loyalist—was someone who was in search of a journey. Not just any journey, mind you, but one that started with a 13,000-mile (20,921 km) trip from her home to Alaska and back in her 1965 Mustang. This was

the Summer of 2017, and her Mustang was 52 years old. She eventually put 500,000 miles (804,672 km) on the car in seven years. That's right, half-a-million miles. Clearly she had been on a number of journeys.

And her time going on such adventures wasn't over. But her time with the Mustang was. The one journey nagging at her was the one Courtney felt was unfinished—and that was to visit the Arctic Circle. For that, she would need a 4×4, and what better vehicle than a Bronco.

Of course, being the individualist she was, her Bronco wasn't going to be just any random vehicle. Like her beloved Mustang, she would find one and rebuild it. The choice was a 1978 model. And its location? Was it anywhere near her home in South Carolina? Of course not. This one was in Wisconsin, necessitating a road trip *before* the road trip.

"It was my first experience with a Bronco," Courtney said. "Honestly I thought it was in a lot better condition than it was. I thought I was going to fix some cosmetic stuff and ended up doing a complete overhaul. The first thing I started on—completely logical—after a terrifying drive home, was the paint," she laughed. "I added my Free Wheelin' stripes, because that's what I do. Sunrise on the ocean—that's the vision I had. Yellow, orange, and white stripes painted on."

Naturally that was the cosmetic approach. The next stage was performance. Courtney, who considers herself a DIYer, was intent on being involved in everything that went into her '78. She installed a 408 stroker with a 351 Windsor block. "I'm a big fan of those motors. My Bronco needed a bit more oomph." She also revised to overdrive from all-time four-wheel-drive, resulting in the change-out of the transfer case, gearing, and transmission.

"When I was in Alaska, participating in this 2,000-mile (3,219 km) loop rally on the Denali Highway, which is all dirt road, I had heard

Every journey begins at step one, in this case when Courtney Barber bought her 1978 Bronco. *Courtesy of Courtney Barber.*

about this other highway that can take you right up to the Arctic Circle, so I knew this was for me."

What's the roadway of choice? Dempster Highway. It's Canada's only all-season public road to cross the Arctic Circle, 458 miles (737 km) of unpaved footprint that covers some of the most beautiful scenery on the planet on the way to the northernmost point in Canada to which one can drive.

There's no stoplights, no traffic jams, and no road rage.

Once Courtney completed her journey, there were many enthusiasts who wanted to hear about it. As a guest on the Broncast, a podcast all about the Ford Bronco hosted by Jonathan Melton of Nashville Bronco and Donnie Whiteman of Dreamweaver Fab, the adventurist had this to say of her travels: "We had no plan for the venture to Arctic Circle. We went up through Montana and traveled via Google. When you're on the road, particularly in the middle of the Arctic Circle—where there isn't a whole lot of help—you really need to know how to fix things." Nevertheless Courtney was quick to emphasize that "People you meet on the road is part of why I love driving and do what I do. Everyone was so polite and friendly, and the surroundings were so gorgeous and breathtaking.

"The favorite part on my Bronco," Courtney continued, "is the hood and the view from the driver's seat. My windshield is a picture frame. It

Once back in South Carolina, Courtney Barber got down to business on Project Road Warrior. With a little help from her friends, she built a Bronco that was not only a reflection of her personality—along with her allegiance to the Ford brand—but also an expression of her desire to exercise her wandering spirit. *Courtesy of Courtney Barber.*

Destination achieved. Through perseverance, determination, and a little luck, Courtney Barber and driving partner Rob Kinnan of *Motor Trend* made their way to the Arctic Circle following the Dempster Highway, Canada's only all-season public road. *Courtesy of Rob Kinnan.*

makes everything look awesome. And of course, the stripes on the hood.

"The trip took 20 days," Courtney concluded. "I ended up at the Woodward Dream Cruise [in Detroit, Michigan, before returning to South Carolina]. Throughout the whole trip, I only lost one headlight, and I brought all these extra parts.

"I think my next trip is to Iceland."

The Process Behind the Restoration

Another woman Bronco enthusiast, Lauren Morris, a.k.a. Bronco Babe Adventures, also has revived a 1978 Bronco and given it a whole new lease on life. She has a YouTube channel where the Bronco's build was chronicled, which was her first experience wrenching.

'Well, I fought every single bolt," she says on one of her videos. "It was a northern truck, so there was a lot of rust."

In a recent interview, Lauren noted that her favorite parts of the restoration process were "first, the community—connecting with them and Bronco lovers, all good people, so supportive and salt of the earth; and two, proving to myself that I can do it—the feeling of satisfaction to start the project, see your work through and then be able to drive it down the road and show it off.

"I changed the hood—love the stripes—half Farrah Fawcett and half Sarah Connor bad-ass chick vibe," Lauren added. "The retro work is stunning and kind of like the Free Wheelin' look.

"It is my absolute favorite when other women love on my Bronco. I'll never forget getting out of my Bronco when I was at Super Celebration [hosted by *Bronco Driver* magazine] and hearing a little girl point and tell her mom, 'See, girls can have Broncos, too!' Yes, sweetheart, we sure can."

Speaking to her involvement with Super Celebration, Lauren noted, "I was blown away by all the people who stopped and talked to me about my Bronco. It was such an incredible experience to meet friends in real life whom I've made through Instagram, and to have people tell me how much they enjoyed watching the progress of my Bronco. It's been a blessing to have this community rooting me on during the process of this restoration, and to meet so many was the highlight of my weekend!"

When not involved with her Bronco, Lauren is a physician dedicated to providing quality health care to her community and likes to say she learned medicine first then learned "mechanical stuff," but the process remains the same—"if someone has a problem, the goal is to provide a technique and help them."

Lauren Morris calls her Bronco *Diana* after Wonder Woman and gets inspiration from enthusiasts and admirers whom she sees every day. She has toured part of the country in her second gen rig, and when the engine wears out, she plans to install a Coyote V-8. *Courtesy of Lauren Morris.*

SEMA and Ford worked together to support the SEMA Businesswomen's Network (SBN), which invited 125 female volunteers who spent more than 5,000 hours installing aftermarket off-road custom pieces on the Wildtrak, a "perfect group effort." The All-Female-Build Bronco debuted at the 2022 SEMA Show. © 2023 Ford Motor Company.

"Whoever designed cars [came up with] a crude version of how God designed us," Lauren speculated. "Same idea, same valves, same pumps. It's just a lot more basic. I have fun working on cars because it translates, just a lot more straightforward."

Going forward with respect to her Bronco improvements, Lauren ponders on the next steps: "What's left? A lot of things I have to do, actually—a vapor-lock issue. An exhaust leak I have to fix. There are 50 things on a list you have to fix.

"It's great to see people get inspired to wrench on their Broncos, and especially great to have women do it. I've inspired more men than women, but always great to see other women get involved. I see so many old Broncos—still kept by owners for utility value or they're sentimental.

More younger men also are getting into the older trucks of all generations; it's my favorite part of the Bronco population: People are passionate about Broncos and understanding what it takes to keep them on the road . . . What it's like to diagnose an issue with the carburetor. We've all been in the trenches, and all been through it together. It's just a great group of people.

"For me," Lauren concludes, "I'll keep having adventures with my present Bronco, checking out so many spots in America and crossing those off my bucket list. When the engine wears out, I'll do a Coyote swap."

It sounds like a wonderful plan.

Speaking of Plans

SEMA, the Specialty Equipment Market Association, has many functions and groups, and one of its most significant, the SEMA Businesswomen's Network (SBN) supported and featured a special Bronco built entirely by a team of women at the 2022 SEMA Show. The Bronco Wildtrack was donated by Ford corporate to SBN for off-road customization and more than 125 female volunteers spent well over 5,000 hours preparing the Bronco for the ultimate adventure.

The upgraded Bronco rides on an ICON Vehicle Dynamics State 8 suspension and rolls on 17 × 8.5-inch (43 × 22 cm) ICON Alloy Thrust satin black wheels surrounded by 37-inch (94 cm) Milestar Patagonia M/T-02 tires. A unique illuminated Bronco grille and custom wrap by Terra Strada Design announced a new exterior, while on the interior, the rear seat was removed and replaced by a small kitchen complete with a Ford Performance 50-quart (47.0L) Classic Series fridge/freezer made by ARB.

Various upgrades to the stock Ford 2.7L (165 cu in) V-6 EcoBoost engine provided more power and smoother exhaust. Taken together, the Bronco highlighted the capabilities of what a team of women could do.

"This was about changing the way the world perceives our capabilities and showcasing this build with everyone at SEMA," stated Haley Keelin, SBN's Volunteer Team Lead. Women ranging in ages and skill sets wrenched on the vehicle, installed the Bronco's roof-top tent as well as Ford roof rack, Yakima platform rack, AMP running boards, and Baja Forged rear bumper. "The perfect group effort," Keelin added.

(continued from page 125)

and entertainment—edutainment, as they call it—which included meals and time in an event-supplied Bronco. Travel to and from the destination and overnight stays were the responsibility of the attendee.

Every marketing pro will advocate, where justifiable, the creation of such an event or activity, called "experiential" in advertising jargon, in which customers participate and get hands-on with the brand. These events can take the form of auto shows, races, driving events, and more commonly, visits to the local dealership. In today's demanding marketplace, where the consumer is pulled in every direction—digital, virtual, and physical—and facing hurdles over which no one has control, such as the pandemic, getting any facetime with a potential buyer is challenging at best. Nevertheless, as an opportunity arises when the customer is in attendance, it is incumbent upon the brand and its personnel to make the customer's experience remarkable and unforgettable, which, in a perfect world, translates into a sale or sustained brand loyalty, or at a minimum, some sort of word-of-mouth, or today, more accurately, "word-of-mouse."

Clearly, part of the goal here is to create a memorable moment. The theme to the message on July 13, 2020, when Bronco sixth gen was first introduced, was to "Celebrate adventure, live life, and find a way to unplug." There's no better way to uphold that mission than to build lifestyle-based programs and related strategies that inspire stronger brand loyalty. That becomes the end game to all this effort. Keeping an owner within the Ford family remains a top priority for all involved parties.

Think about the length every dealer goes to reach out to its customers and how much Ford the manufacturer spends, both nationally and regionally, in support to its individual dealers. Additionally the aftermarket spends funds of its own to attract targeted audiences after their vehicle purchase to encourage them to upgrade their vehicle in the name of personalization, customization, and utilization. This aggressive marketing approach is separate although frequently complementary to the original equipment manufacturer (OEM) as these parts, equipment, and services are in most cases dictated by the consumer's budget and want, but not need, unless they are accessories such as tonneau covers, tool boxes, winches, and larger tires that may be critical for one's business. It's an evolution.

SENSE OF BELONGING

To welcome new members and further encourage a sense of community, enthusiasts of Bronco and Blue Oval have developed multiple clubs that have spread throughout the United States, Canada, and worldwide. Certainly one of the oldest Bronco-only clubs is the Bronco Club of America. Taking a page from the successful Mustang Club of America (MCA) and calling itself a grassroots site, the Bronco Club states it is "designed to provide historical data, facts, and inspiration for this important and growing community." Naturally the organization quickly built a sizable audience early on.

Dave Landwehr's Bronco 6Gen (www.bronco6g. com), which came to life during the early discussions about when Ford would build a new Bronco model, saw membership quickly grow from 80 enthusiasts—"many of them Dearborn people, designers and builders of the new Bronco"—to 75,000 "within two years." Clearly the insiders who formed the foundation brought in many more willing participants.

Various forums have functioned for years, though many have moved either to a Facebook or Instagram page or a more aggressive website that includes event planning and offers merchandise.

Bronco Nation is an incredible source of information for Bronco enthusiasts, with news, videos, and forums, and is currently the only nationally recognized and certified Bronco community by Ford. Organized by Jackson-Dawson Communications, a creative marketing agency based in Dearborn, Michigan, more than 15,000 members are regularly entertained and informed through YouTube videos hosted by Jordan Parker and Laura Zielinski, and thousands more participate in various Bronco Nation-supervised forums.

Parker has become the de facto face of Bronco Nation and regularly travels to all of its events as well as other Bronco programs and says his goal is "to build upon what we call the Bronco community and make sure participants in our events get the most out of their vehicle.

"Off-roading generally has been one specific group of people," Parker continued, "and some are intimidated about trying it, or they feel they don't fit in with that group. But Bronco Nation, and all the folks connected to it, wants to get new people to join in, and we're seeing more women especially coming forward, as well as folks who may have thought they were too old—nonsense. It's all about being active . . . redefining or rediscovering a passion for the

Club identity, as well as badge envy, is part of the fun in bringing people and their vehicles together. *Images courtesy of Bronco Nation; sixth-gen badge image courtesy of TK1215 Life is a Highway.*

OPPOSITE: The Off-Roadeo program—a Bronco-owner-only event held four times per year at selected locations around the United States—allows owners and those who have one on order to drive a Ford-provided Bronco off-road with instruction from experienced drivers. It's one way Ford connects directly with its customers. © *2023 Ford Motor Company.*

Bronco Nation, Ford, and Jackson-Dawson all had a hand in creating *The Legend Returns*, a short film produced in Iceland chronicling the triumphant return of the sixth-gen Bronco to a land where its first gen siblings sold quite well. Iceland turned out to be a very receptive market for both the old and new Bronco and may be a good bellwether when Ford begins sales of Bronco sixth gen to Europe in late 2023. *Courtesy of Hooniverse.com.*

outdoors or a new hobby in a new vehicle. Things have really changed postpandemic, the smiles are large as people use their Broncos as a way to connect and Bronco Nation is the richer for it. It's very special to me."

Parker also served as executive producer of the film *The Legend Returns* that was filmed in Iceland, which interestingly, was a huge market for Bronco first gen, with 2,000 units sold there. Parker pitched the idea of having a sixth gen visit the land its ancestors conquered 45 years earlier, and it was approved by Ford. He planned the whole route and handled the logistics of shipping a film crew as well as vehicles from Detroit (through Portland, Maine) to Reykjavík, Iceland.

"It was seven 16-hour days on ferries, off-road, and on-road and the most rewarding experiences and connections of my life," Parker exclaimed. "We were able to organize a whole meet-up with early Bronco owners and it was amazing. The passion and general appreciation of Bronco remains strong no matter where you are. The entire experience added a new spin on the whole Bronco perspective for me. I think we as consumers are thankful Ford has this vehicle that's not only approachable but allows more people to get out there on the road or off the road in whatever comfort zone they want. Most importantly Bronco builds a community and relationships who share these experiences and I'm proud to part of that."

Among other sites, there's also classicbroncos.com that covers early Broncos, offers more than 600 links to respectable sources, a garage for vehicle display, tech articles, and a vendor forum.

And to be more specific, a Bronco enthusiast has many choices and forums including the following: Full Size Bronco (fullsizebronco.com); Blue Oval Forums (blueovalforums.com); Ford Bronco Forums (fordforums.com); Bronco Corral (broncocorral.com); Ford Truck Enthusiasts (ford-trucks.com); Bronco Zone (broncozone.com); Bronco Raptor Forum (broncoraptor.com and broncoraptorforum.com); Four Door Bronco (fourdoorbronco.com); and Early Bronco Registry (earlybronco.com). Nor Cal Broncos (norcalbroncos.com) offers up one as well and no doubt other regional clubs have their own as well as Ford sibling brands such as Ranger, Maverick, F-150, and other automotive forums with Bronco as a topic du jour or ongoing discussion.

One additional Facebook page that has seen increased support over the past year as the number of owners has expanded is Ford Bronco 2021+ (but not Bronco Sport), which is filled with valid questions, humor, and enthusiastic members.

PRESS ATTENTION

Naturally Bronco has earned its share of press coverage throughout the years, obviously a goal of Ford's media and marketing personnel, whose combined efforts get Bronco stories in magazines, on television, internet video, Twitter, blogs, and newspapers. The goal is to make Bronco "top-of-mind" among journalists and ultimately, consumers, because that leads to recognition, desire, and sales.

Magazine covers, feature stories in the Holy Grail of newspapers, *The Wall Street Journal*, and pieces on programs such as *NBC Nightly News* are the focus because they get worldwide attention and essentially a "third-party endorsement" when an impartial reporter submits a positive review.

The "big four" automotive consumer publications, *Road & Track*, *Car and Driver*, *Motor Trend*, and *AutoWeek*, have paid considerable attention to Bronco over its life and especially since its relaunch. *Car and Driver* recognized the 2023 Bronco in its 10 Best Trucks and SUVs list for the second consecutive year.

Bronco Driver is the one single-purpose publication that has been around since 2012 and has logically gathered steam since the return of Bronco. *Courtesy of Bronco Driver magazine and Profiles Marketing Group.*

Additionally niche publications such as *Vintage Truck* and *Hemmings Classic Car* have featured Broncos on their covers and appeal to their reader base, sort of like preaching to the converted. *Vintage Truck's* editor and publisher, Brad Bowling, has an affinity for Ford products, as he and the author worked together in the late 1980s for Saleen Autosport, a producer of limited-edition Mustangs.

There's one magazine that stands out in the world of Broncos because it's first and only focus is, well, Broncos. Tom and Donna Broberg started *Bronco Driver* before the reboot of Bronco and appreciatively captured the stories of the first five generations, but with the new vehicles, got an abundance of fresh material. It continues to grow as a result.

ADDITIONAL APPRECIATION

Ford earned double honors for two years in a row from the North American Car, Utility and Truck awards committee, taking home the 2022 North American Utility of the Year award for Bronco and the 2022 North American Truck of the Year award for its Maverick pickup. (Ford won in 2021 for its Mustang Mach-E in the Utility category and F-150 in Truck. The F-150 Lightning took the award for Truck in 2023.)

North American Car, Utility and Truck of the Year (NACTOY) awards, formerly announced during the January North American International Auto Show but still given during that month, are decided upon by a distinguished panel of 50 jurors, recognizing the most outstanding new vehicles of the year: those that define or redefine their segments with superlative attention to innovation, design, safety, handling, driver satisfaction, and overall value.

Another award earned by Bronco, perhaps more coveted because it was close to home, was recognition from *Detroit Free Press* by winning the 2022 SUV of the Year (with family member Maverick taking first place as 2022 Truck of the Year).

And those are just a few key moments—and recognition—in the history and life of Bronco.

9   In media reports, no greater investor than Warren Buffett, whose company Berkshire Hathaway owns 5.5%, has said that Apple has developed an ecosystem and level of brand loyalty that provides it with a "competitive moat," and consumers appear to have a degree of price insensitivity when it comes to the iPhone.

10   Corbin Reiff (November 8, 2014). "20 Years Ago: The Eagles Release 'Hell Freezes Over.'" *Ultimate Classic Rock.*

The 2022 Bronco won the North American Utility Vehicle of the Year, part of the North American Car, Utility and Truck of the Year (NACTOY), that year at the North American International Auto Show in Detroit, an important recognition and honor for the new SUV. © 2023 Ford Motor Company.

BRING BACK BRONCO

THE UNTOLD STORY

The eight-part *Bring Back Bronco: The Untold Story* podcast is hosted by veteran NPR auto reporter Sonari Glinton. The popular podcast, produced by Pacific Content, tracked the rise, untimely fall, and welcome rebirth of the iconic Bronco. © 2023 Ford Motor Company.

Chapter 11

The Customer Speaks OUT LOUD

" Your Bronco is a beautiful example of machinery, technology, and quality, produced by humans and robots in a united ballet of beauty and purpose, more than one thousand times a day."

It might not exactly be something Ford would send a customer, but the image is.

In fact, Ford CEO Jim Farley—who takes customer engagement very seriously, especially at driving events and racetracks, which fuel his own motorsports passions—has stated that customers with confirmed orders were receiving photos of their vehicle at the Michigan Assembly Plant and would continue doing so as the vehicles were completed on the assembly line.

Ford continues to ramp up its involvement with order-placing Bronco customers beyond its Bronco Off-Roadeo program, by sending or offering posters, toy models, hammocks, and other items to buyers waiting for their Bronco. New Outer Banks owner Robin Calef Kirtlan wrote on the Ford Bronco 2021+ (not Bronco Sport) Facebook page that she had been waiting two years and one month since placing her order but when her Bronco arrived, all had been forgiven, she was so happy to have it. Now, she wrote, the personalization would begin.

PODCASTS

Worldwide, podcasts have grabbed listeners' attention in increasing fashion because of the format's success in establishing a virtual one-on-one with what has become a truly engaged audience. And unlike focusing on videos, podcasts, like radio, are "theater for the mind," allowing the listener to participate without a screen. Many take in a podcast when driving to and from work, like one used to do with radio, or from any location, for that matter.

Additionally sources such as entrepreneur.com say podcasts that appeal to a specific individual allow that listener to become a part of a community of like-minded people and is beneficial to brands. Vocast.com indicates "research shows that *people who listen to podcasts are more focused on the content as they have actively chosen to listen to the specific episode. As compared to other media, listeners don't jump from one podcast to another.*"[11]

Among those within the Bronco community, there are several that stand out:

Ford Bronco Talk (a.k.a. FoBroCo), started in September 2020, is owned by TruckTalkMedia and features a number of different people behind the mic including Drew Paroni, early Bronco specialist; Todd Zuercher, early Bronco specialist and historian; Julian Carr, a.k.a. "Blake Bronco," second-gen specialist; and Cliff Brumsfield, second-gen specialist. Truck Talk Media Owner and Producer Ronnie Wetch also hosts from time to time and is the editor of all shows, now numbering more than 56 podcasts dedicated to Broncos. The company also offers F-100 Talk and for listeners of other marques, C10 Talk, OBS (Chevy/GMC), and the Dodge Pod.

Donnie Whiteman and Jon Melton cohost *The Broncast*, which obviously is all about the Ford Bronco, attracting 23,600 subscribers produced by Nashville Early Bronco YouTube channel. Both hosts earn their livings working on Broncos in various capacities. Jon builds custom Broncos.

Another popular Apple podcast, *Bring Back Bronco: The Untold Story*, focuses on how Bronco found its place again and features host Sonari Glinton, who was an assembly line worker at Ford, along with his mother and seven others in his family.

In its initial write-up, the eight-part podcast series asserted, "Join us on this wild ride . . . an 8-part serial uncovering 50 years of blood, sweat, and dirt."

The details within Glinton's podcast cover a myriad of subjects; they are pop culture, history, car design, and the decision-making among numerous attempts to regain approval of the Bronco behind the scenes. Glinton also has worked as an automotive reporter for National Public Radio and gained a strong following for his stories about money, jazz, and award-winning business coverage. He has contributed to *The New York Times Magazine*, *The Advocate*, and the BBC.

Ford sent photos of customers' completed Broncos to let them know their vehicle was in process in terms of its journey from Michigan to its new home. This one belongs to customer Robin Calef Kirtlan in California. © 2023 Ford Motor Company and Ms. Kirtlan.

Bronco podcasts attract many listeners and cover valuable topics. *Ford Bronco Talk* is one popular example. *Logo courtesy of Ronnie and Autumn Wetch, TruckTalkMedia.com.*

Other podcasts available include the following:
- #broncotalk
- #broncopodcast
- #fordbroncotalk
- #broncoholic
- #broncitis
- #fordbroncopodcast
- #FoBroCo
- #BroncoNation
- #60sBronco
- #EarlyBroncoNation
- #EarlyBroncoCrew
- #4×4#keeponbronco'n
- #4×4Bronco#BroncoDriver
- #BroncoDriverMagazinze
- #dentsidebronco
- #78_79Bronco
- #BroncoClubofAmerica
- #newBronco
- #GenSixBronco
- #BroncoCommunity
- #BroncosofInstagram
- #classicbronco
- #VintageBronco

THE IMPACT OF BRONCO ON POPULAR CULTURE

"Hey, wasn't that a Bronco in that shot?"

Ford Archivist Ted Ryan has a cool job, especially for anyone who loves history, and Ford Motor Company, at 120-years-old, has a lot of stories to preserve, as well as rely upon, when planning for its future and its enduring legacy.

Ryan is charged with keeping Ford's heritage alive and robust. Throughout Bronco's life and during his lengthy career at the Blue Oval, the archivist has collected more than 1,000 films, TV shows, songs, and other references that include the iconic Bronco. One would think people would run out of ideas or have forgotten the vehicle that last appeared as showroom-new in 1996. But with age comes grace, coupled with collectability and that "cool factor" that advertising and marketing types talk about. Long before the new Bronco appeared, its predecessor portrayed that sense of self-assurance, which transferred to all who drove one.

As soon as the new Bronco came out, the coolness went up several notches. It's a go-to vehicle, and artists have gravitated toward it, positioning Bronco as a centerpiece to any story. It's the same for those posting on Facebook and Instagram. Bronco is a part of our lives.

Music

It's unsurprising that a majority of the songs Ted Ryan has found thus far feature country music artists. He's discovered at least 35 songs or music videos, 2 hours and 10 minutes of music, with some either obvious or oblique reference, or visualization, to Bronco.

This is a list in no particular order:
- *She Get Me High* by Luke Bryan
- *Daddy Made the Dollars, Mamma Made the Sense* by Doug Supernaw
- *Birdman Kicked My Ass* by Wesley Willis
- *Bronco* by Canaan Smith
- *May We All* by Florida Georgia Line, Tim McGraw
- *Same Old Song* by Blake Shelton
- *Cosmic Cowboy (part 1)* by Nitty Gritty Dirt Band
- *Genuine Article* by Kid Rock
- *Flipmode Enemy #1* by Rampage
- *My Ol' Bronco* by Luke Bryan
- *Back in the Saddle* by Aerosmith
- *Don't Matter* by Akon
- *Drive (for Daddy Gene)* by Alan Jackson
- *We are Tonight* by Billy Currington
- *Bubbly* by Colbie Caillat
- *Isn't That Everything* by Danielle Peck
- *Lovin' You Is Fun* by Easton Corbin
- *Cruise* by Florida Georgia Line
- *Hold Me* by Jamie Grace and TobyMac
- *Stay Gone* by Jimmy Wayne
- *Up All Night* by Jon Pardi
- *Nothin' to Lose* by Josh Gracin
- *My Life Would Suck Without You* by Kelly Clarkson
- *How Country Feels* by Randy Houser
- *Runnin' Outta Moonlight* by Randy Houser
- *Take Me There* by Rascal Flatts
- *By the Way* by Red Hot Chili Peppers
- *So* by Static-X
- *The One* by Tamar Braxton
- *Heartbeat* by The Frey
- *Don't Cha* by The Pussycat Dolls and Busta Rhymes
- *Big Love* by Tracy Byrd
- *That Don't Sound Like You* by Lee Brice
- *Confident* by Demi Lovato

The good folks at *Bronco Driver*, Tom and Donna Broberg, have not only built a beautiful publication, which is no easy trick to do in today's digital-based society, but also fill it with excellent content and wonderful stories. One of the sections is called "Word on the Trail," where they list new Bronco-based songs and other activities that feature any of the six generations.

The list of songs is a never-ending process and will need updating as time goes on. Here are a couple of additional songs, courtesy of Tom:

- *Beat up Bronco* by Leah Turner
- *You Ain't Going Nowhere* by Jake Owen

Film & TV

For Clint Eastwood fans, a first-gen Bronco played a key role in the actor's 1975 thriller film about mountain climbing, *The Eiger Sanction*. Filming took place in most of 1974, so there's some argument as to whether it's a 1974 or 1975 customized Baja Bronco. It features many of Stroppe's

special equipment and package, but has the wrong interior for a Baja, according to Andrew Norton, who runs Baja Broncos Unlimited, as well as many participants in a 2004 forum on the film on classicbroncos.com.

There were some notes to the fact that Stroppe rented the vehicles (there were two: one dedicated to filming and one as a support) to Malpaso Productions, Eastwood's production company, but the company never returned one of them, asking "how much?" and sending a check for the amount Stroppe wanted. Rumor suggested Eastwood was the original buyer and then someone else from the film crew later bought it.

Clint Eastwood plays art professor and mountaineer Jonathan Hemlock in 1975's *The Eiger Sanction*, which he directed. The Baja Bronco enjoyed a major role in the film that was produced by the Malpaso Company and distributed by Universal Pictures. *Alamy Stock Photo.*

Here's a selected list of other famous films with appearances by Broncos:

- *127 Hours* (2010)
- *13 Going on 30* (2004)
- *2 Guns* (2013)
- *Ace Venture: Pet Detective* (1994)
- *Always Be My Maybe* (2019)
- *Ambulance* (2022)
- *American Gangster* (2007)
- *American Pie 2* (2001)
- *Arachnophia* (1990)
- *Bad Boys* (1995)
- *Bad Boys II* (2003)
- *Basic Instinct* (1992)
- *Body Snatchers* (1993)
- *Brainstorm* (1983)
- *The Breakfast Club* (1985)
- *A Bronx Tale* (1993)
- *Bumblebee* (2018)
- *Charlie's Angels* ★film reboot (2000)
- *The China Syndrome* (1979)
- *Dog* (2022)
- *E.T. The Extra-Terrestrial* (1982)
- *The Eiger Sanction* (1975)
- *Fast & Furious Presents: Hobbs & Shaw* (2019)
- *Fear and Loathing in Las Vegas* (1998)
- *Free Willy* (1993)
- *The Hangover* (2009)
- *Hope Floats* (1998)
- *Independence Day* (1996)
- *Interstellar* (2014)
- *Iron Man 2* (2010)
- *The Italian Job* (2003)
- *Jaws 2* (1978)
- *The Klansman* (1974)
- *The Last Song* (2010)
- *The Lincoln Lawyer* (2011)
- *Logan* (2017)
- *Man of the House* (2005)
- *Marley & Me* (2008)
- *Men in Black* (1997)
- *Narrow Margin* (1990)
- *Need for Speed* (2014)
- *A Nightmare on Elm Street* (2010)
- *No Country for Old Men* (2007)
- *North Dallas Forty* (1979)
- *Ocean's Eleven* (2001)
- *Octopussy* (1983)
- *Office Space* (1999)
- *Orange County* (2002)
- *Quantum of Solace* (2008)
- *A Quiet Place: Part II* (2020)
- *Rain Man* (1988)
- *Rampage* (2018)
- *Role Models* (2008)
- *Romancing the Stone* (1984)
- *She's in Portland* (2020)
- *Sleepless in Seattle* (1993)
- *Smokey and the Bandit* (1977)
- *Speed* (1994)
- *Swimfan* (2002)
- *Terminator 2: Judgment Day* (1991)
- *Three Days of the Condor* (1975)
- *Top Gun: Maverick* (2022)
- *Uncle Buck* (1989)
- *Walking Tall: The Payback* (2007)
- *Zodiac* (2007)
- *Zoolander* (2001)

Most recently, a four-door red Bronco Wildtrak with custom lighting had a nice supporting role in a new ABC show *Not Dead Yet*.

Actor Jason Segel is a car person, and his new show for Apple TV+, *Shrinking*, features a vintage Bronco that his character drives throughout the first season. Segel drives a vintage Bronco in real life, so he wanted to have one in his show.

Also, a Bronco first-gen was featured in the NFL-produced intro to the 2023 AFC Championship game, "A Child's Game."

And Bronco Sport had a nice end-of-episode one appearance in the season-opener of Netflix' new spy thriller, *The Night Agent*, as well as brief entrances in later installments.

There are literally hundreds more in films and on virtually every popular TV show from the mid-1960s to current day, as well as multiple TV commercials, mostly very quick sightings, some with the main stars/characters driving or riding in them and others where Bronco simply makes a cameo. Either way, it's definitely a vehicle that's always camera-friendly and clearly enjoys the limelight.

BRONCOS IN MOVIES

From IMCDB, the Internet Movie Cars Database (www.imcdb.org/vehicles), Ford Archivist Ted Ryan assembled another 1,890 obscure films and mainly TV shows, including international productions, that have some Bronco inclusion within, either static or moving. It's a pretty amazing list and took a great amount of time to develop, considering one had to find and view each one of these and locate the Bronco within. Lack of space here prevents listing every program, but here is a selection of the author's favorites as well as several recent shows the author has viewed and added.

TV SERIES	YEAR	SERIES SPAN
Alias	1969 Wagon	2001–2006
Better Call Saul	1994 XLT	2015–2022
Big Sky	1996 XLT	2020–present
Bionic Woman	1970 Sport top removed	1976–1978
CHiPs	1966 Roadster	1977–1983
Dallas	1982	1978–1991
Deputy	1979	2020
Entourage	1977 Wagon top removed	2004–2011
Green Acres	1967 Roadster	1965–1971
Hart to Hart	1981	1979–1984
Hawaii 5-0 (multiple shows)	1966 Wagon, 1970 top removed	1968–1980
Justified	1989 Eddie Bauer	2010–2015
Longmire	1994 XLT	2012–2017
Magnum, P.I.	1984 Bronco II	1980–1988
Mannix	1967	1967–1975
Mare of Easttown	1st gen	2021
Miami Vice	1978	1984–1989
Monster Garage	1968 Pickup	2002–2006
Narcos	1982	2015–2017
Narcos: Mexico	1982	2018–2021
Northern Exposure	1970 Wagon	1990–1995
Police Story	1968	1973–1987
Six Feet Under	1982	2001–2005
Sons of Anarchy	1988 Eddie Bauer	2008–2014
The Rockford Files	1978	1974–1980
The Rookie: Feds	1st gen	2022–2023
The X Files	1978 Ranger XLT Free Wheelin'	1993–2002
Twin Peaks	1989 XLT	1990–1991
Walker, Texas Ranger	1970	1993–2001

WOULD YOU BUY A CAR FROM THIS MAN?

Then, there's the story of John Bronco.

John is (was?) a good ol' boy in the best way possible: rugged, handlebar mustache, not really a cowboy, but more *cowboyesque*. He seemingly displays a rodeo background but unknown, part Western outdoorsman, part mystery man, with a cool demeanor. Could he be considered a sex symbol like Burt Reynolds on the bearskin rug in *Cosmopolitan* magazine (1972)? That's doubtful.

Nevertheless, he fit a certain mold, and Ford snapped him up as a pitchman, so the lore states. He was featured in a number of ads for the original Bronco, but as time went on and John's ego possibly got a bit fried under the Sun, things began to go south.

There's a short film on Mr. Bronco's brief life as a figure-head for the first waves of Bronco, but just like the music video that destroyed rocker Billy Squier's career (if you saw it, you would wholeheartedly agree), the shorts John wears in one scene are sure to put a dagger in his future career as a promotional spokesperson. In fact, the big mystery surrounding John Bronco is that he disappeared in 1996, ironically, about the same time the fifth-gen Bronco was put out to pasture. Perhaps he was also.

Of course, this is all done with tongue squarely in cheek. *John Bronco* the film features Walton Goggins in the title role, a multitalented actor who can play tough guy, as in the series *Justified*, as easily as a widowed dad in *The Unicorn*. What the 40-minute "mockumentary" focuses on is "an affectionate ode to America's love affair with big cars, bigger advertising and the outdated macho cowboy image," according to *Detroit Free Press* reporter Julie Hinds. It also has been hailed as an "instant classic" in terms of its comedic overtones toward how a new vehicle would logically be launched.

Ford played along. The film, scheduled for release on the streaming service Hulu on October 15, 2020, was right at the time the automaker was dealing with its own launch of the new Bronco, and had it not seen the value in supporting such a film would have missed the opportunity for some viral and incremental exposure for its lineup. In fact, when the *John Bronco* production team approached Ford's public relations folks, the latter opened up its archives and helped find vintage vehicles used in the film. The director, Jake Szymanski, was even allowed inside the company headquarters.

Still, in the end, John Bronco goes missing. Bo Derek also stars in the film. If John were smart, he'd probably have run off into the sunset with her. In a Bronco, of course.

Famed Ford salesman John Bronco had it all: a Wild West personality, persuasive skills, and a smile brighter than the lights of Las Vegas. Too bad he was a fictional character. But it made for a fun story and good publicity for Ford. *Courtesy of Hulu and Imagine Documentaries.*

Imagine Documentaries and its Executive Vice President of Brands for Imagine Entertainment, Marc Gilbar, a Bronco fan himself, served as producer of the 37-minute "mockumentary" *John Bronco*, released in 2020 on Hulu. *Courtesy of Hulu and Imagine Documentaries.*

A BRONCO OWNER'S LOVE LETTER

Transportation authority Dave Kunz, KABC-TV's long-time automotive reporter (in Los Angeles) and cohost of "The Car Show" on KPFK-FM, owner of a 1977 Bronco Ranger, explains how his affection for his vehicle and the marque has done nothing but grow in the 18 years since he purchased it.

"After owning a couple of classic Mustangs for a number of years, in 2005, I decided I might like to have one of the first-generation Broncos I'd always admired," Kunz started. "After a months-long search for an uncut, non lifted example, I hit paydirt.

"The family of a well-known Bronco collector in Colorado was selling his beloved 1977 Ranger model after his passing. One of the latter ones off the Michigan Truck assembly line before the changeover to the redesigned 1978 models, it was sold new in Chino Hills, California, about 45 miles (72 km) from my home. It was fully-loaded, and the dealer had also equipped it with air conditioning, as it was not offered from the factory. The prior owner had purchased it from the original owner in the early 1990s."

Kunz continued: "When I first went looking for an early Bronco in 2005, you could still buy a pretty nice one for $20,000 or less. In the ensuing years, collectors across a wide spectrum have made the Bronco a bit of a brass ring in the classic truck world. I've been tempted to consider offers, but I'd probably regret selling it almost immediately. My wife would also veto such a move, as she enjoys owning it as much as I do!

"This '77 Bronco Ranger was bone stock when I first bought it, with the exception of an aftermarket four-point roll bar. After a few years driving it that way, I decided to go for a 'Day 2' look with nonpermanent period modifications. Basically, the things an owner back in the day would have added via accessory catalogs like the one by legendary Bronco racer Bill Stroppe . . . a set of 15-inch (38 cm) aluminum 'slot' wheels with B. F. Goodrich All-Terrain tires; vintage Cibie off-road lights; a smaller three-spoke steering wheel; and a column-mounted AutoMeter tachometer.

"With its uplevel—for the time—Ranger interior, 3.50 gears, and the more street-friendly later—'76 to '77—steering linkage and a stock ride height, the Bronco actually doesn't do too badly with highway travel," Kunz noted.

"I added extra sound deadening, a modern sound system, and in the summer the A/C keeps the interior nice and comfortable, helped by the added tint on the rear windows. Factory dual fuel tanks also give it a pretty decent range of 250 miles (402 km) or more."

The perfect companion. Always a great ride.

BRONCO FEVER + BRONCO ACTIVISTS = "BRONCOLAND"

Finally, nothing is more important than what the consumer says (or thinks), and that's reflected in the way they express themselves through words and actions. A list of vehicles from several folks who responded to the author's request are displayed here along with commentary from active participants in several Bronco clubs. Here we go:

"We try to make it to Super Celebration every year, a couple car shows, and parades," the native of Oklahoma said. Obviously a great time is had by all.

Speaking of events, late 2022 saw the first of its kind occur in Arizona. Called the Arizona Bronco Round-Up and organized by the Arizona Bronco Club and Bronco Nation, it was held December 17 at Sanderson Ford in Glendale, AZ, with approximately 200 Broncos from all six generations represented. All agreed it was a great time that included such luminaries as Willie Stroppe, Baja Broncos' Andrew Norton, Shelby Hall and Todd Zuercher, author of the original Ford Bronco book and highly regarded for both his passion for Bronco and his technical prowess, as well as many others who were there to represent the Bronco community.

"We will return to Sanderson Ford this December," said Joey Bombaci, one of the event organizers, "and this time, we're expecting 400 Broncos."

This is Jack Vest's beautiful 2021 Black Diamond two-door in Cactus Gray. Jack's president of the Alabama Bronco Society, a moderator for Bronco 6th Gen's Facebook page, and passionate for all things Bronco. "My love for Bronco radiated through my social media posts, which gave me the opportunity to connect with Bronco enthusiasts across the world," Jack added. "The sixth-gen Bronco gives owners the ability to have the best of both worlds . . . whether you want to take your family to church on Sunday or spend the weekend hitting the trails, our Bronco is the perfect vehicle to do just that.

"The moment Ford teased the idea of bringing back an icon, I knew I had to have a sixth gen," Jack enthused. "I made my reservation the night the order banks opened. Little did I know that it would be 466 days before my dream became a reality, but it was worth the wait. Keep on Bronc'n."

Bronc'n is what people do. The following images are from the various Facebook and forums mentioned previously and help to define both customization and personalization and the Bronco ownership spirit.

Bobby Tennell's September 1965-built 1966 U14 Half Cab is "in original overall about 99% original" including its Arcadian Blue paint and original upholstery. "We try to make it to Super Celebration every year, a couple car shows, and parades," the Oklahoma native said. *Courtesy of the owner.*

Dave from Bend, Oregon, (known as "yippeekiyay" on social media forums) owns a 2021 Badlands Sasquatch two-door here parked in an appropriate setting. *Courtesy of owner and courtesy of Bronco6gen.com.*

Auto Nation illustrator (and physician) Alexander Nicholson owned this Bronco when he lived in Hawaii. Here it receives special attention from a dedicated (and cheap) detailing professional. *Courtesy of owner.*

Imagine being the owner of both a Bronco and a Bronco-branded camper shell. This 1971 Bronco 3-speed was bid to $59,000 on Bring a Trailer, but failed to meet reserve. *Courtesy of owner.*

THEY'RE NOT VANITY PLATES, THEY'RE VALIDITY PLATES

Pictures tell a thousands words, so enjoy a sampling of the passionate and expressive.

11 https://vocast.com/the-power-of-podcasting

Ford installed everything but the kitchen sink on this two-door Bronco. Inside it features an ARB refrigerator/freezer, recovery/assistance/first aid kit, center console vault, and red trimmed door handles. On the exterior, this Bronco wears a Ford Outfitters off-road Free Wheelin' Package, a WARN Industries winch, RIGID Industries Light Bar, and Yakima-supplied pieces, including a Hi-Lift Jack Bracket. To prove its long-distance capabilities, this Ford Outfitters Bronco wears the Sasquatch Package wheels and tires, along with recovery boards, tube doors, CURT hitch shackle mount, and red *Bronco* lettering on the grille.

Chapter 12

The Aftermarket
SPEAKS

I t's amazing how many people are involved and literally earn their livings, or even a portion of it, through their involvement with Bronco. It's said that approximately 10 percent of the U.S. workforce (16 to 64) takes home a paycheck from some job within the automotive industry, from a service station to an auto repair shop to motorsports to marketing. Think about the manufacturers, their dealerships, the agencies that serve them, the suppliers, tire companies, oil businesses, transportation services, Amazon and so-called "last mile" deliveries—the list is endless. Add to that the electric vehicle (EV) eruption of the last six, seven years, and the number of people employed keeps moving upward.

The total market revenue of the automotive manufacturing (factories alone) business is roughly $104.1 billion. The entire industry generated approximately $2.86 *trillion* in revenue in 2021, the last numbers available, and as the pandemic began to wear away, that market size was expected to grow another 3.3 percent in 2023.[12]

And then there's the aftermarket. The Specialty Equipment Market Association (SEMA) based in Orange County, California, and producer of the annual SEMA Show, the second largest trade show to descend upon Las Vegas, claims in its 2023 SEMA Market Report companies providing products and services to consumers postpurchase of their vehicles is now a staggering $51.8 billion industry and is expected to continue to grow 4 to 5 percent per year.

Obviously consumers, buyers, of automotive products want to personalize, individualize, and customize their own vehicles, and the industry has successfully responded to their needs and wants. For every fully loaded First Edition Bronco, there is still something else its owner desires to make it "theirs." And there's absolutely nothing wrong with that—nothing at all.

A PEEK AT A FEW BRONCO-CENTRIC BUSINESSES

Let's take a look at several companies that specialize in business of all things Bronco. So many companies draw their revenue from some aspect of the Bronco world. This is a call-out on a number of them.

Builders

The following is a brief list of professionals who are dedicated to building, restoring, and customizing the esteemed Bronco.

Firehouse Vintage Vehicles

In Statesville, North Carolina, sits a small business known as Firehouse Vintage Vehicles. The company specializes in classic Bronco and truck sales and restorations. With 50 years of combined experience, the Firehouse team, led by founder Justin Surprenant, can claim credit for a number of restored Broncos, particularly second gen, as well as F-Series trucks. The company completed its first 5.0L (302 cu in) Coyote swap on a second gen in 2018 and in 2022, installed its first 7.3L (445 cu in) Godzilla-powered engine in a lucky 1978 second gen in Candyapple Red over Wimbledon White.

Firehouse lists a number of "Ready to Roll" Broncos and Ford trucks for sale on its website as well as several "Barn Finds & Builders," ones that could use some work and sell for how they appear and what they're worth.

Velocity Modern Classics

For Stuart Wilson and co-owner Brand Segers, Velocity Modern Classics of Pensacola, Florida, is single minded in how they view vehicles people use: "Travel is a state of mind." As a result, it isn't necessarily the journey, but what you drive in (or in what you drive). The company specializes in modernizing 1966–1977 Broncos as well as International Scouts and a wide range of vintage vehicles, embracing the core values of each subject's history and design, while prioritizing quality, luxury, and performance during the build process. Led by Wilson, a Navy nuclear machinist vet, precision is the name of the game, and since 2019, Velocity has experienced tremendous demand, now producing 120 luxury Classic Ford Broncos each year.

Vintage Broncos

Perpetually a serial entrepreneur, Chau Nguyen has created a business that he's into "for life." A car fanatic since birth, he recognized Bronco first gen as a "beautiful, timeless vehicle."

"The Bronco started the SUV craze," Nguyen declared, whose shop, Vintage Broncos, sits outside Atlanta in Buford, Georgia, "and it is one of the most sentimental vehicles in history. Being a 50+-year-old SUV, there's no way you haven't been in one or at least around one of the originals. People want an early Bronco because of that emotional attachment."

For Velocity Modern Classics of Pensacola, Florida, "Travel is a state of mind." That mindset encompasses an owner who is willing to invest significantly in a modernized Bronco first gen. This 1973 Velocity Modern Classic is an example of the quality of the company's work. *Courtesy of Velocity Modern Classics.*

Vintage Broncos of Buford, Georgia, specializes in fully restored first gen Broncos fitted with a Coyote V-8, 10-speed automatic, dual Bilstein shocks, A/C, power steps, windows, and steering, Wilwood six-piston disc brakes front and rear, custom upholstered leather interior, and other appealing and functional modifications worthy of a $239,000 asking price. *Courtesy of Vintage Broncos.*

Because "vintage is in style, people who can afford a restored Bronco want to relive being around that vehicle," Nguyen added. "However, the whole thesis behind my business is that because the Bronco is more than 50 years old, nothing really works. So my job is to make sure everything does. We keep it vintage but make it modern."

Nguyen emphasizes that he sources everything for the customer and hand builds each Bronco to spec, doing whatever the customer wants. He noted that his company has stayed true to its roots by just focusing on first generations, and with the dedication to each build, "it makes spending $200,000 on a car acceptable. And everyone knows the value of an early Bronco, from start to finish."

Maxlider Brothers Customs

The best way to describe this builder is "mind-blowing." Maxlider Brothers Customs definitely deliver unusual Broncos for people who favor the unique. A six-wheeled version is among the most noteworthy, but one cannot turn away from the company's four-door gen, its version of the "vintage Bronco."

The Bloomington, Illinois, operation is run by Erik Barnlund, a.k.a. Erik Maxlider, who serves as CEO, and his brother Kris, whose title is CTO for chief technology officer and the one who manages the shop, having spent 16 years

at a Ford dealership where he earned all ASI Senior Master Ford Tech certifications.

The brothers' goals are straightforward: to help their customers create stories with their families that last a lifetime by delivering the kind of transportation they want. Whether it's a finished project vehicle, a fully restored, show quality, high-end custom Bronco, what Maxlider calls a "vintage Bronco reimagined," such as the four-door version, Coyote conversions, sixth gen upgrades as well as 2021+ Ford F-150 and Ranger mods, the Maxlider brothers will answer the call.

ICON

Although Detroiters will argue vehemently, Los Angeles is for many the automotive capital of the world, and it's here where Jonathan and Jamie Ward run ICON. The incredible success of the custom builds, best known as restomods, is based on "the actualized results of a strong commitment to tradition, obsession with modern design and unrelenting need to achieve performance excellence," Jonathan noted. It is this focus on vehicles with classic styling, modern performance, and timeless utility that renders fantastic values of its numbered Broncos.

ABOVE: Though the Bloomington, Illinois, business is called Maxlider Brothers Customs, perhaps it should really be known as Maxxlider, the double *x* connotation referencing the Maxlider brothers' incredible six-wheeled Broncos. *Courtesy of Maxlider Brothers Customs.*

Los Angeles-based ICON is a builder of high-end Bronco first gens. ICON BR 101 was sold on Bring a Trailer in January 2023 for $291,000. This *Derelict* model is so-named because it retains its exterior patina earned over a long life. Modernizing features include 18-inch (46 cm) alloy wheels, a Fox Racing suspension, Brembo brakes, heated front bucket seats, and an Alpine touchscreen infotainment system. VintageAir A/C, fire extinguisher, ICON vintage interior knobs, and other details complete the look, while the fuel-injected 5.0L (305 cu in) Coyote V-8 provides the power. *Courtesy of BaT and Buchholz of Five Points, South Carolina.*

Classic Ford Broncos (CFB) offers "bespoke" (custom-made) trucks through a manufacturing process that owner Bryan Rood says isn't just about building a Bronco—it's about "building experiences." The company starts with a Ford-licensed factory-reproduction tub and builds everything from the ground up. Customers can specify a 302 or Coyote V-8. This Chrome Yellow and California Cream 1975 Bronco features a Gen 3 Coyote 5.0L (305 cu in) V-8 and is called *Ocean Boulevard*. It's available from CFB for $249,900. *Courtesy of Classic Ford Broncos.*

Although not their first line of vehicles—those were Land Cruisers—the Wards eventually turned to Bronco first gen and have introduced their incredible ICON BR, which can be configured in a full hardtop, soft top, half cab hardtop, or Roadster. The ICON brand is emblazoned across the reconstituted Bronco grille. ICON BRs also are quickly recognized for their "Old School ICON wheels," their dedicated identity number, Coyote engines, and other signature accessories and trim, making them both highly prized and extremely desirable vehicles.

The company reached its "iconic" number 100 milestone in 2022, a significant accomplishment given the fact that ICON competes in a tight marketplace of similar high-quality manufacturers, including those listed here. Each company offers a unique vehicle, a customized automotive piece of art, and as ICON's Jonathan Ward maintains, "an authentic driving experience—a blending of new and old." That statement is no doubt appreciated by ICON's loyal buyers and owners, as BR 101 recently sold on Bring a Trailer for $291,000.

Classic Ford Broncos

"Sometimes it seems like half of the trucks I sell end up in the Hamptons," said Bryan Rood, owner of Classic Ford Broncos, a custom builder based in Powell, Ohio. *"People there get used to seeing the same luxury cars over and over, but they may not remember the last time they saw an early Bronco."*

Bryan Rood of Powell, Ohio, pours his passion into his work, and it clearly shows. He finds first gens wherever he can and gives them a "21st-century makeover," an approach the craftsman believes sustains the classic look of the original Bronco and then improves it with modern mechanicals. Customers who wish to order a custom-built Classic Bronco have a list of choices and options such as a blueprinted 302 or a Gen3 Coyote, which would include a factory-original 1966–1977 Bronco frame and licensed Ford factory reproduction tub.

Additional tech upgrades make the restored Classic Bronco more than capable of performing well on roadways and off-road. Meticulousness is a key word when it comes to the work Rood and his team performs, as "restoring vintage cars is a dying art," Rood notes. The entry point for one of these classics is $179,000, so clearly they are works of art.

Engine Factors

What is the appeal of a Coyote engine?

When properly tuned, the 5.0L (302 cu in) engine offers 435 hp (320 kW) and 400 lb-ft (542 N·m) of torque. This highly regarded engine is offered in the Mustang GT, and when installed within a lighter-weight, restored Bronco, makes for a much improved, namely faster, ride. And these significantly modified vehicles who carry one in their engine bays are certainly worth the value they're assessed.

And then there's the Hennessey VelociRaptor 400 Bronco. It was created by engineer extraordinaire John Hennessey, who earned a reputation for building vehicles that exceeded any normal speed limit, especially 1000 hp (736 kW) sports cars, and his company, Hennessey Performance, which counts more than 12,000 specialty vehicles produced, the new VelociRaptor 400 and 500 Bronco were announced in 2022 and 2023, respectively.

The Hennessey VelociRaptor 400 Bronco joins a long line of Hennessey Special Vehicles offered to customers who want a unique fast-moving SUV with emphasis on the *sports* portion. *Courtesy of Hennessey Special Vehicles.*

The engine choice for this modified Bronco was an upgraded twin-turbo 2.7L (165 cu in) V-6 of 411 hp (302 kW) and a whopping 602 lb-ft (816 N·m) of torque, propelling the Hennessey SUV from 0 to 60 miles per hour or 0 to 97 kilometers per hour in 4.9 seconds. Meanwhile, the VelociRaptor Bronco 500 is based on the 3.0L (183 cu in) V-6 and packs 500 bhp and 550 lb-ft (746 N·m) of torque. An off-road package is an option on both variants, making these among the fastest multipurpose SUVs available.

One cannot ignore the other side of the motivational (as in power) coin either—an EV conversion. While not a common occurrence, at least two companies are offering them, Kincer Chassis and Gateway Bronco, with EV components from Legacy EV.

Kincer Chassis of Louisville, Tennessee, is well known for its custom first-gen chassis fabrication and restoration. Kincer will build customers everything from a new frame to a complete rolling chassis and perform a full restoration, depending on the individual's request and budget. In 2019 owner Thomas Kincer was asked to build an EV Bronco and feature specs from first gen's launch year of 1966. The goal was to design a new chassis platform that would properly support all battery boxes and EV components and keep them hidden from sight as well as protected. The result was a fully electrified first-gen Bronco that was just as capable as any internal combustion engine–managed vehicle while embracing its original aesthetics and design.

Legacy EV provides education, advice, and many choices when it comes to the proper EV power supply. It supported Gateway Bronco on the latter's build of its LUXE GT Bronco first gen. *Courtesy of Kincer Chassis.*

Legacy EV of Tempe, Arizona, prides itself on providing solutions to those seeking to convert their vehicle to EV power as well as educational support, components, and build advice. The company has advised on many aftermarket EV applications, helping companies considering modifying their fleets into electrified vehicles and providing support to automotive shops as they transition to providing service to electric-powered vehicles as well as selling EV conversion kits.

In addition, the company is extremely proud of its role in the build of Gateway Bronco's Luxe GT Electric Bronco that features a 1969 Bronco powered by a Danfoss Motor and a custom 106 kWh battery pack producing 295 hp (217 kW) and an incredible 861 lb-ft (1,167 N·m) of torque.

Gateway, based in Hamel, Illinois, also custom builds Broncos and recently partnered with Roush to produce a Roush Performance Bronco, which Gateway is calling "the world's first generation 3 Coyote with 10-speed automatic and Roush supercharger" (the same as in the new Shelby GT500) placed inside a Bronco first gen.

Service, Support, and Extras

The following businesses and enthusiasts provide the Bronco customer with premium parts, accessories, information, updates, and most importantly, a sense of community.

Fox Factory

The "Ride Dynamics" producer based in Duluth, Georgia, known for its racing shocks and other performance equipment, has partnered with off-road legends Jason Scherer and Dave Cole to offer enthusiasts a special KOH Bronco. KOH stands for the annual Ultra4-produced King of the Hammers off-road race held on the rough trails in Johnson Valley, California, and is typically the start of the desert racing season and one of the toughest events of the year.

Scherer, a multiple time KOH and Ultra4 Champion and 20-year off-road racer, took a Bronco sixth gen and designed it to be lightweight and agile off-road while equally adept as a daily driver. Fox already has a relationship with Bronco as the supplier to Bronco Raptor with technology enabling it to endure harsh terrains. The KOH Bronco is built to produce a more comfortable ride. Bronco's Wildtrak trim uses Fox' position-sensitive tech for fast desert runs, while the KOH version features Fox 2.5 Performance Elite shocks with dual-speed compression adjusters for versatility and tuneability. Special *KOH* badging and interior design reflect the uniqueness of this vehicle, and Fox and Scherer are working with individual Ford dealerships to enable customers to order directly from them.

Wild Horses Four Wheel Drive

Jim Creel, owner of Wild Horses 4×4 of Lodi, California, has a long history of supporting the Bronco community, offering high-quality parts and products for all generations. Nearly 35 years of service to Bronco owners has encouraged him to continually expand and move his operation, finally consolidating at an old Ford dealership, giving him the space he needed to set up his R&D shop and launch the annual "Bronctoberfest" Swap Meet, a popular event much of the Bronco community attends. Creel and his crew also organize the Wild Horses Bronco Roundup, a two-day show and off-road event that runs in May with more than 250 participants.

CJ Pony Parts

Harrisburg, Pennsylvania, is home to CJ Pony Parts, primarily known for the past 38 years for its Mustang parts and accessories. However, as a loyal supporter to Ford enthusiasts, Jay Ziegler, one of the original founders who still runs the company, recognizes that many Mustang owners have another Blue Oval vehicle in their garage, and it's probably either a truck or a Bronco. Thus, the company wisely expanded into carrying parts for the F-100, F-150, and Bronco first gen, and now sixth gen, further expanding its customer base. The aftermarket business also produces a popular DIY installation series of videos found on YouTube.

Bronco Driver Magazine

In an era where regrettably, magazines are a dying breed, it's heartwarming to see *Bronco Driver* succeed and grow. Thanks to the sixth gen, new life in "Broncoland" means more stories, more readers, and subscribers and most importantly, new companies serving Bronco interests who become advertisers. Tom and Donna Broberg have been publishing *Bronco Driver* since 2002, after an extensive career in marketing and advertising where many of their clients required them to produce publications such as city and tourist magazines and even catalogs. Being car people since birth and with their blood completely Ford blue, thanks to Donna's father's passion for Ford trucks and Broncos and

A logical brand extension for *Bronco Driver* magazine is event management, which has become quite successful since the 2006 launch of the company's Super Celebrations. With the relaunch of Bronco, the company has experienced a growth spurt with additional events introduced. Originally started in Tennessee, they have added events in Colorado, Wisconsin, and Nevada. *Courtesy of* Bronco Driver *magazine and Profiles Marketing Group.*

Tom's interest in Mustangs, they soon discovered that a magazine about Broncos was a valuable niche and in April 2002, launched issue number one.

The company's Super Celebration started in 2006 when the F-100 Super Nationals asked if *Bronco Driver* could "get a few Broncos to its event since it was the marque's 40th anniversary," according to Tom. "We thought we could get a dozen or so out for the weekend, but after 200 came piling in through the doors, it was apparent we might have a market for a stand-alone Bronco show." From 2006, a Super Cel was held in Tennessee, and since that time, the Brobergs have added shows in Colorado and Wisconsin and in 2023, a new event in Carson City, Nevada. Ford has most recently used these events to unveil Bronco sixth gen and the Raptor, and this year will set up Ford Base Camps at each.

Bronco Driver has a bright future, and the Brobergs are looking to organize what they are calling "Bronco Campfire Cruises" in the near future. "These will be limited space events," Tom said. "One- and two-night overlanding and destination drives for Bronco enthusiasts wanting to get out and experience more with their rigs. It will be open to all generations of Broncos." And speaking of generations, today, Tom and Donna drive a '76 Bronco. Their daughter Brittany (named after Tom's 1967 Brittany Blue Mustang convertible) owns a '77. Their sons Brett and Bryan both drive Broncos too: Brett with a '92 and Bryan in a '76. Eight grandchildren round out the family tree, and all are named after cars. "The cycle of building and buying kids Broncos *may* continue," Tom noted. *"May?"*—most likely.

FordAuthority.com

He could be considered the "hardest working man in journalism," tracking down information on Ford (and other marques through a series of additional online news services that he oversees), including everything from in-house memos and rumors to patent applications to factory upgrades, and Alex Luft, founder of the news source, shows no signs of slowing down. An automotive enthusiast at heart, Luft has a strong passion for anything on wheels and is the Founder and managing partner of Detroit-based Motrolix and its properties, of which Ford Authority is one. Brett Foote handles the on-deck writing duties for the most part.

(continued on page 156)

Ask virtually anyone who has gotten their toe wet in the automotive buying and selling business about Bring a Trailer (BaT) at bringatrailer.com (now owned by Hearst, which also owns *Car and Driver* and *Road & Track* as well as manages Autoweek.com, among many other publications, websites, and newspapers and to which the author has no affiliation), and undoubtedly, there will be a nod followed by a quick comment, "Did you see how much that [insert vintage vehicle here] sold for today?"

Yes, some of the sales prices occurring on this extremely successful auction-based website are incredible. The final amounts result from value, ego, rarity, presentation, ego, lottery winners and one percenters, desirability, "buying for one's wife," need, and oh, did we mention ego?

Founded in 2007 by Randy Nonnenberg, who serves as chief executive, and Gentry Underwood, BaT is a daily obsession for many (author included), attracting more than two million visitors per month. It's amazing the assortment of vehicles that have appeared on the site, but even more so the number of such heralded beauties that live in collectors' garages. It seems as though every Jaguar XKE convertible ever made has been listed—at least once. Trends are quickly established. If one or two vehicles from one marque sell well, more will soon appear. The old phrase "strike while the iron is hot" always rings true when it comes to rising values on cars or just about anything

that becomes popular and subsequently takes on a new perception.

Celebrating its 100,000th auction recently, BaT started to climb in popularity before the pandemic, but as more people had time on their hands—and money burning in their wallets, apparently—when COVID-19 hit, prices on many highly desirable vehicles began to creep upwards and then rapidly, approaching heights only previously seen at in-person, high-end auctions. The commentary that consistently runs throughout each auction, many of which borders on Jerry Seinfeld's (who's been both a buyer and a seller) level of storytelling, in addition to offering fabled expertise, wisdom, and experience, is what drives the value of many of those vehicles listed, not to mention the audience, as each offers both advice and general well-intentioned criticism when needed.

To Bronco fans, what's been pure joy and sheer surprise (some might add disbelief and/or distress, depending on whether or not they sold their own too early) is the vast number of both first and current generation models listed and sold at amazing numbers. Even more incredible are the early models modified by ICON and other companies that command staggering sums. Most recently, the later generations are getting in on the action and selling through at stunning figures.

Here are five recent examples of Broncos selling at big numbers on BaT, with thanks to BaT for utilizing some of the copy from these examples. Images approved by the sellers.

1973 Ford Bronco Explorer

A 1973 Ford Bronco Explorer was acquired by the owner in June 2020 and subsequently underwent a significant refurbishment that included installing a 302 cu in (5.0L) V-8, four-speed AOD automatic transmission, two-inch (5 cm) suspension lift kit, and a one-inch (3 cm) body lift as well as replacing and repainting the entire body, reupholstering the interior in black vinyl, and adding a roll bar and a powerful sound system. Finished in light green, the truck is equipped with a dual-range transfer case, a Dana 44 front axle, a 9-inch (23 cm) rear axle, a removable hardtop, polished 17-inch (43 cm) US Mags wheels, a quick-remove door hinge kit, and electric windshield wiper kit. A total of 81 miles (130 km) have been run since the transformation occurred.

1974 Ford Bronco Ranger

A 1974 Bronco which reflected a significant refurbishment completed in 2022 that included repainting in the orange and cream (creamsicle?) and an overhaul of the replacement 302 cu in (5.0L) V-8 engine and three-speed auto transmission and dual-range transfer case. Additional work included 17-inch (43 cm) cream-colored wheels, 2.5-inch (6 cm) suspension lift, front disc brakes, gas rear shocks, plus tan vinyl upholstery and houndstooth inserts with matching door panels.

Of course, some of these are outliers or unicorns, but it's a tribute to the seller and the buyer who recognize quality and value. As most commentators and veteran owners of such historic vehicles—and any vintage car or truck, for that matter recognize—the cost of restoration can very quickly approach high five and even six figures based on the detail of that renewal. We're not even going to talk about ICON Broncos here. The author can't count that high. (There is some additional information on ICON in this chapter.)

1977 Ford Bronco Sport

A 1977 Ford Bronco is finished in blue over Parchment knitted vinyl upholstery—with matching dashboard and door panels—and dispenses power from a 302 cu in (5.0L) V-8 mated to a three-speed automatic transmission, a two-speed transfer case, and a limited-slip rear axle. Equipment includes a removable white hardtop, 15-inch (38 cm) steel wheels, dual exhaust outlets, swing-away tire carrier, Sport Bronco Package, auxiliary fuel tank, heated front seats, center console, air conditioning, wood trim, and a RetroSound radio.

1978 Ford Bronco Ranger XLT

A 1978 Bronco XLT Ranger, the first year of Bronco's second generation, became the seller's vehicle in 2021 and was repainted in the original Candyapple Red and Wimbledon White, a popular two-tone combo. Brakes and suspension were replaced and a Wieand intake manifold installed along with an Edelbrock four-barrel carburetor, electronic ignition, and dual exhaust, tied to the optional 400 cu in (6.6L) V-8 available during the second generation, along with a three-speed auto transmission and dual-range transfer case. Additional features include the split front and folding rear bench seats upholstered in red vinyl and patterned cloth epitomizing the fashion of the 1970s.

Tilt steering column, cruise control, AM/FM, swing-away spare tire carrier, and matching white steel wheels mounted on 33-inch (84 cm) rubber give this Bronco a beefy look, supported by power steering, power-assisted front disc brakes and rear drums, manually locking front hubs, and dual front shocks on each side, which were replaced, along with front coil springs.

No Reserve: 38k-Mile 1995 Ford Bronco XLT

The final featured Bronco is a 1995 XLT which remained with the original owner until 2022. Finished in Deep Forest Green Clearcoat Metallic and Light Opal over Opal Gray cloth upholstery, it's powered by the desirable 5.8L (351 cu in) V-8 linked to a four-speed automatic transmission and dual-range transfer case. It was equipped with automatically locking front hubs, stock 15-inch (38 cm) wheels, chrome bumpers and trim, power tailgate window, A/C, cruise, power windows and door locks, and CD stereo. At only 38,000 miles (61,155 km), this fifth gen was barely used compared to many of its siblings that have had multiple owners and have rolled their odometer at least once.

Strangely this generation has not appreciated in value as rapidly as the first—which no others have come close to in any event, though second gen is starting to move upward (see below).

Of course, the response to all these online auction buys is best expressed as, "Why not?" It's a case of supply and demand . . . or aptly put "beauty is in the eye of the beholder." . . . You have something I want, and I have the means and the ways to acquire it. Simple as that. It works at in-person auctions such as Mecum and Barrett-Jackson too, though the premiums are typically higher.

As one winning bidder noted, "Bought my Bronco months ago—1975—in pristine condition for $35K. Thought I was crazy. These cars are classic and fun. Well worth the investment."

Not to be morbid, but the discussion quickly turns to YOLO (you only live once—unless, of course, you're Shirley MacLaine) in justifying the purchase just made.

In a recent post on LinkedIn, the adult version of Instagram, Allen Aguilar included a quote from actor Denzel Washington, which is quite appropriate to this train of thought:

"You'll never see a U-Haul behind a hearse . . . I can't take it with me, and neither can you. It's not how much you have but what you do with what you have."[13]

If only that were the case all the time. Then everyone would have a bevy of Broncos in their garages.

Jake Gertsch is a true Bronco fan, owner, and businessperson, as well as a devoted Montanan, having smartly combined all four into his company Montana Broncos. *Courtesy of Jake Gertsch.*

Hella Bad Broncos, the Austin, Texas–based proprietor of United by Bronco, began by offering first- and second-gen rebuild and repairs, but it has rapidly grown into providing full frame-off and original restorations, including sixth gen. This *Hellazilla* is an example of the team's incomparable work. It's seeing snow for the first time while being trailered to a show. *Courtesy of Hella Bad Broncos.*

(continued from page 153)

Montana Broncos

To Jake Gertsch, owner of Montana Broncos in Billings, Montana, "the community is what makes it worth owning a Bronco. It's hard to explain the iconic culture, but once you experience it, you get hooked," he said. "Obviously to experience it, you buy a Bronco. Or go to an event. It's amazing how friendships are forged over a brand. This vehicle defines a community—Broncos are amazing in that way—people everywhere have their own individual cars and clubs, but everyone will compliment a Bronco when it comes into a parking lot."

Gertsch should know, he's currently got 15 Broncos—"currently, at least one from each generation"—in his collection and has owned somewhere around 50, and he's still young. He buys and sells Broncos, performs upgrades to them, and is well known for his Bronco Stripes, which are colorful side-long applications or hood-mounted decals primarily for sixth gen and second gen.

"Retro is in, and stripes are in," Gertsch stated. "I created a stripe package with a different take on it than the factory. I wanted a factory tri-stripe for my restored Bronco, so I created one and then took it to market for the second gen and subsequently started with the newest gen. Athletes and celebrities want these stripes on their vehicles. It's giving new life to the second gen. Two years ago, you wouldn't see a '78 or '79 at Barrett-Jackson, but this year there were seven. Says something about demand.

"Living in Montana is a great place to enjoy adventures in the outdoors, and Ford built the Bronco for just that," he continued. "Broncos will always be part of the family."

United by Bronco

This group of like-minded individuals, led by Jesse Ornelas of Hella Bad Broncos—with Bronco obviously at the core—have organized two events where owners from every generation unite for numerous off-road adventures, most recently in March 2023 at Sand Hollow Park in Hurricane, Utah. Numerous sponsors and attendees participated in the three-day event that for the first time included a street cruise. With new presenting sponsor DV8, United by Bronco will return again to Hurricane in 2024, "loving life," as Ornelas put it.

The Austin, Texas–based Hella Bad Broncos business started as a Bronco first and second gen rebuild and repair shop that grew into full frame-off restorations and has now added service as well as upgrades and modifications to Bronco sixth gen along with lowered and lifted F-100s and other makes.

Ford Aftermarket Parts

Recognizing that there is tremendous potential in parts sales, particularly in aftermarket upgrades to Bronco sixth gen, and especially since there are now more than 550,000 new Broncos in circulation, Ford has stepped up its game and inventory. The Blue Oval has developed many new upgrades available to Bronco customers postpurchase that can be acquired from local dealerships.

Additionally Ford has partnered with aftermarket manufacturers to produce parts and accessories on its behalf. In fact OEM-sanctioned aftermarket parts sales have increased by 40 percent in the last two years with gains felt by corporate as well as its dealer network. More than one million factory-endorsed Bronco accessories have sold. Accordingly Ford recognizes that advocating vehicle customization is an important part of its business today.

Part of this plan was clearly developed in conjunction with the original sixth-gen design, adopting a modular concept that enables the doors, fenders, and grille to be removed and replaced with alternatives, making off-road "tailoring" easy and quick for enthusiasts.

"Whether your goal is to build the ultimate desert racer or rock crawling rig—or both with the same vehicle—the Bronco modular design provides the ease and confidence to create a 4×4 that is as unique as each of our customers," said Bronco Chief Designer Paul Wraith. "Because of that ease and flexibility, no two Broncos should ever be alike."

Bronco logos stamped on the body and interior fasteners indicate removeable modular components and unlike competitive SUVs, both front and rear fenders can be removed by simply taking out the bolts and replaced just the same without welding or other complicated attachments.

Threaded Bronco logo mounting points from the factory enable the addition of accessories like LED pod lights to sideview mirrors and installation of roof racks to the top sport bar. Available modular front and rear bumpers also come with threaded taps to mount winches, safari bars, and more LED lights.

"We designed Ford vehicles to be easier to customize, with installation points and aftermarket accessories engineered from the start of the design process," echoed Eric Cin, Ford Global Director, Vehicle Personalization, Accessories and Licensing. "We work actively with the aftermarket to share our product designs so customers can personalize with the finishing touches on their vehicle to make it all their own.

"Owners will keep their Bronco fresh and up to date with new body parts, accessories and new technologies— some that can be made in low volume and others downloaded," said Wraith. "The possibilities for the new Bronco platform stretch as far as your imagination."

The commencement of Built Wild was set.

This heavy-duty Bronco Badlands four-door built in partnership with RTR Vehicles sports a number of custom aftermarket accessories such as an RTR Signature Grille with LED Lighting, 35-inch tires and RTR Tech-6 wheels and special RTR graphics that complete the package. © 2023 Ford Motor Company.

SEMA Show Broncos

In what was an annual tradition for the Blue Oval at the biggest automotive aftermarket show in the world, SEMA, and to underscore the goal of increasing parts sales, special show vehicles built either by Ford or with the automaker's support were featured in both Ford's booth at the front of Central Hall or within the booths of major vendors and equipment manufacturers.

In 2021 Bronco's first year, the Ford booth displayed a number of sixth-gen Bronco and Sport show vehicles, ranging from mild to wild, demonstrating what each model is capable of delivering, depending on budget and designation.

Of course, the Ford booth wasn't the only exhibitor to show off Broncos. They were featured throughout the show, which was regaining its footing as the pandemic wore down. SEMA officials were pleased with the turnout and with multiple Bronco appearances, named it the SEMA 2020 4×4/SUV of the show, which it clearly deserved.

Ironically Ford did not exhibit at the 2022 SEMA Show, choosing to have other companies feature its vehicles and special Bronco builds within their respective booths. Despite the lack of the Blue Oval's physical participation, it remained in Las Vegas in spirit. And the market remains strong for aftermarket parts in any event.

12 https://www.zippia.com/advice/automotive-industry-statistics/

13 https://www.linkedin.com/pulse/20140829140718-329335609-remember-you-can-t-take-your-money-with-you-when-you-die/

Here are some of Bronco's SEMA highlights since the launch of the sixth gen:

TOP LEFT: Ford Performance Parts uses the annual SEMA Show to launch both prototypes and examples of its customized packages. The show has featured both the Bronco and Bronco Sport since their introductions in 2021. For SEMA 2023, Ford showcased this Blue Free Wheeling Bronco–created through Ford's Accessory Personalization concept package–highlighting components and graphics, making customers' vehicles everything they need for surf and sand. © 2023 Ford Motor Company.

TOP RIGHT: A regular builder of Ford specialty vehicles, BDS Suspensions, a subsidiary of Fox Shocks, offered up an off-road first-responder based on a 2021 two-door Black Diamond Bronco, equipped with a BDS 4-inch (10 cm) UCA System, Fox 2.5 PES coilovers, BDS rear adjustable control arms and track, and belly skid plate plus recessed winch mount and recovery hooks front and rear. An ARB Twin Air Compressor sits underhood. Courtesy of the author.

ABOVE: This is a modern-day version of the Sno-Cat seen in the movie version of Stephen King's The Shining. Tucci Hot Rods built this quad-track "snow adventure vehicle" based on a 2021 four-door Badlands. The show-car veteran used eight-series tracks in place of tires and fitted its intrepid warrior with Ford Licensed Accessories such as swing gate storage, flat snowboard rack, and Yakima LockNLoad platform roof rack system. No snow rig of this caliber would be complete without a RIGID Industries Adapt Light Bar, Radiance Pods, and Rock Light Kit. A WARN Industries winch and retractable running boards complete the package for the ultimate snow-bound excursion. Courtesy of the author.

TOP LEFT: Sara Morosan, co-owner of LGE-CTS Motorsports, created this rough-and-ready 2021 Baja Forged Bronco Sport Badlands with equipment from Ford Performance Parts and Ford Licensed Accessories, including a 2-inch (5 cm) suspension lift and rock sliders. The exhaust note of the 2.0L (122 cu in) EcoBoost engine is enhanced with a Borla Cat Back exhaust system. Suspension enhancements include ICON Vehicle Dynamics suspension system and Hellwig Suspension Products front and rear sway bars. This Baja Forged Bronco Sport wears *BAJA FORGED* front and rear tubular bumpers, plus a Ford Performance Parts roof mounted light bar. A WARN Industries winch keeps it on the trails. *Courtesy of the author.*

TOP RIGHT: This 2022 Badlands Bronco Sport custom built by Yakima and Hypertech was designed to give drivers the ability to extend their off-road journeys. The SUV features an ARB portable air compressor, exterior and underbody lightning, and Aeroskin hood protector. Additionally Ford Performance Parts provided a rooftop off-road lightbar and Borla exhaust, while Yakima and Hypertech provided a rooftop tent and shower for the ultimate in off-road adventure. *Courtesy of Yakima/Hypertech.*

ABOVE: ARB's Service Unit 2022 Wildtrak Bronco is inspired by the rugged, efficient medical service transporters of the past. Those emergency-response vehicles of yesteryear are given a modern upgrade with tube doors, an underhood air compressor, winch, and light bar to ensure that this Bronco custom can provide assistance wherever it is summoned. *Courtesy of ARB.*

The official unveiling of the new Bronco was not at a
dealership or an auto show, but in the desert, at a race,
at the best place, Baja, to be exact. © 2023 Ford Motor
Company.

Chapter 13

Racing
Becomes
REALITY

Putting your money where your mouth is. There's no better, stronger, more justifiable proving ground than the racetrack, or rather, the competition arena, be it asphalt, cement, dirt, rocks, mud, or any other surface (in the case of earthbound vehicles).

When Ford decided to introduce its first Bronco, the manufacturer took it to Baja.

When Ford decided to introduce its sixth-gen Bronco, the manufacturer took it where?

To Baja, of course.

And, certainly, for those who love motorsports and off-roading, what better way to prove Bronco's four-wheel drive capability but to race it. In terms of highlighting key moments in the life of Bronco, one has to be the fact that when the vehicle was first launched in 1966, Ford used the Baja 500 as an opportunity to showcase its ruggedness, and did so again in 2020, when Ford Performance ran a Bronco R Race Prototype in the 53rd SCORE International Baja

1000. Piloted by Cameron Steele, Shelby Hall, and veteran Ford off-road racers Johnny Campbell, Curt LeDuc, and Jason Scherer, the entry in the November 2020 event in Class 2 added to Bronco's storied history in the Mexican landscape.

STEPPING BACK IN TIME

In partnership with race car builder Bill Stroppe, who worked closely with Holman & Moody, the new 1966 Broncos also competed in the Mint 400 and Mexican 1000 (which became the Baja 1000). The relationship continued and Stroppe/Holman & Moody entered six Broncos in the Baja 1000 in 1969. In 1971 a Baja Bronco Package was made available through Ford dealers, which offered many of the features of the off-road racer, including quick-ratio power steering, fender flares, roll bar, reinforced bumpers, padded steering wheel, and a distinctive red, white, blue, and black paint scheme. They also had automatic transmissions, which may have dampened sales a bit. Approximately 500 (some

Rod Hall and Larry Minor helped build the kind of credibility that Ford needed to prove its Bronco was a G.O.A.T. (goes over all terrain) when the two famously drove Bronco to an overall win more than 50 years ago at the 1969 NORRA Baja 1000. © *2023 Ford Motor Company.*

It must have been a sensational feeling, firing up the Bill Stroppe-prepared Bronco for an assault on the NORRA Baja 1000 in 1969. © *2023 Ford Motor Company.*

sources say 650) were sold over the next four years, priced $1,900 above a standard V-8 Bronco. Today remaining Baja Broncos are worth six figures, according to Andrew Norton, who maintains the actual production number is closer to 450 from the full run during 1971 to 1975, and has made a career restoring and modifying these rare and wonderful off-roaders.

Norton has always been a Ford fan, buying a Mustang when he was 14. But all bets were off when he saw his first Baja Bronco two years later. When he found out it was a Stroppe Bronco, that's when he started his plan to go into old car restoration. Eventually that very same Baja Bronco he'd seen as a teenager became his.

"I opened Baja Broncos Unlimited in 1998," he started. "I did research on the trucks, interviewed Stroppe employees, read everything I could and became an authority on them, and turned that into a business."

Business in his Petaluma, California, shop has been good, and Norton has had many long-time customers. As a purist, his work takes time—he's done seven full restorations, using as many new old stock (NOS) parts as he can find—and shows no signs of stopping. "When I find a Baja Bronco that has survived, some have been taken care of, some are sitting in a barn, wrecked, the difference between the types of vehicles is like knowing which have been raced

and which have been chased—the latter by collectors like me, who either know their value or don't, and sometimes I get lucky and can buy one or help the owner by bringing this piece of history back to life."

Fast forward to 2020, and Bronco, not yet ready for prime time, was prepared to go racing.

ON YOUR MARK . . .

"When Bronco returned, we said it would follow in the legacy of the first-generation Broncos that forever changed the off-road landscape," noted Mark Rushbrook, Global Director, Ford Performance Motorsports. "[Participating in Baja] . . . demonstrates we're continuing the 'Built Wild' pedigree of Bronco."

Powered by a fully stock Ford 2.7L (165 cu in) EcoBoost engine and Ford's 10-speed SelectShift automatic transmission, the Bronco R Race Prototype served its purpose in a "trial-by-fire" assessment for Bronco Built Wild Extreme Durability Testing and for the final validation of the Baja mode calibrations for the Terrain Management System within the model's G.O.A.T. modes on the production version.

A preproduction 2021 Bronco two-door Outer Banks model with Sasquatch off-road package also made its first appearance in the Mexico's Baja, California, desert.

The 2020 Bronco R Race Prototype in action at Baja. Getting its feet wet is not really accurate—more like getting its shocks absorbed or getting its sand on? © *2023 Ford Motor Company.*

The successful finish that saw the Bronco R finish the 53rd Baja 1000 in Class 2 in just over 32 hours underscored Bronco brand's performance heritage in Mexico, which includes the first-ever overall production 4×4 class win in the 1969 Baja 1000 by a stock Bronco, driven by racing legends Rod Hall and Larry Minor—a feat no other manufacturer has accomplished since, according to Ford's media personnel.

Overall the first-gen Broncos claimed five Baja 1000 class wins (1967, 1968, 1969, 1971, and 1972) and two overall Baja 500 victories in 1970 and 1973. Later model year Broncos (1978–1995) continued the legendary winning streak with 9 Baja 500 Class 3 wins from 2004 to 2015 and 15 Baja 1000 Class 3 wins between 2002 to 2019.

And even before the new Bronco R's debut at Baja, Bronco Brand Ambassador Shelby Hall, Rod Hall's granddaughter, and navigator Penny Dale drove a Bronco Sport to win the 2020 Rebelle Rally X-Cross class.

The Rebelle Rally would shine brightly in Bronco sixth gen's win column in subsequent years, including a three-peat

Before it saw a spec of sand, the design of the new Bronco R was intended to display an extreme version of the new Bronco, to incorporate Ford's "Built Wild" DNA, and showcase the stock SUV's High-Performance Off-Road Stability Suspension or HOSS, System. And it had to be impressive, which it is. © *2023 Ford Motor Company.*

A perfect pairing. Old and new Bronco meet on familiar territory. © 2023 Ford Motor Company.

The Rebelle Rally is an all-women competition that covers more than 1,200 miles (1,931 km), beginning near Lake Tahoe, Nevada, and ending at Imperial Sand Dunes in California. It's the ideal terrain to prove the capability and durability of the Bronco Sport small SUV. © 2023 Ford Motor Company.

at the 2022 event in the X-Cross Class where Bronco Sport piloted by Melissa Clark, a Bronco Off-Roadeo Moab Trail Guide (and 2021 X-Cross winner), and navigator Chris Benzie, also a rally vet, would take first.

Shelby Hall and Penny Dale competed in the 4×4 class in the same rally in their two-door Wildtrak Bronco with a HOSS 3.0 System Package upgrade and earned sixth.

"I specifically requested our two door Bronco because of how classic it looks and how much it resembles the 1968 Bronco my Papa (Rod Hall) used to race," said Hall. "The livery is what Ford calls the race fleet's war paint, and the colors are derived from my grandfather's winning 1968 Bronco.

"This Bronco speaks volumes of the passing of the torch," she added. "The Ford Bronco history started with my grandfather, and now continues with me."

Ford, however, was ready for more, and in late 2021, announced that it would take its desert racing efforts to a new level with a special Bronco DR, a limited-edition Baja 1000–dedicated race vehicle based on its successful sixth gen.

Developed and built in partnership with Multimatic, the same Canadian firm that helped engineer and produce the Ford GT supercar, Bronco DR is essentially a turnkey desert-racing machine and includes a full roll cage, additional suspension travel for high-speed jumps, and a 5.0L (302 cu in) V-8, similar to that found in the Mustang GT and Mach 1, developed to produce more than 400 horsepower (298 kW). Its purpose is for off-road use only and engineered to put enthusiasts behind the wheel of a "desert racing force," according to Ford's Mark Rushbrook.

The "Built Wild" mantra used liberally throughout the Blue Oval definitely came true with the Bronco DR. With a race weight of 6,200 pounds (2,812 kg), its approach angle of 47 degrees and departure angle of 37 degrees with a breakover angle of 33 degrees give the off-roader capability to handle severe tracks. A 73.7-inch (187.2 cm) front track

plus a 73.3-inch (186.2 cm) rear track provide secure tire patches on the surface. In addition, large air intakes on the upper takes and a massive scoop on the roof help absorb air to cool the rear-mounted radiator. A 65-gallon (246L) fuel tank rides below the cargo tank allowing the DR to run long and fast.

Bronco DR retains the four-door model's frame and body structure underneath the fiberglass body panels. Side panels replace doors while the face maintains the brand's signature grille and production round headlamps.

What's even more amazing is that when Ford announced in late 2022 that it was going to build 50 production versions at close to $300,000, no one blinked an eye and the limited-edition model sold out quickly. That says a lot for the faith, trust, and respect buyers and enthusiasts have in this incredible racing machine.

Shelby Hall and Penny Dale partnered again for the 2021 and 2022 Rebelle Rally running a two-door Bronco in the 4×4 class and took fourth and sixth place, respectively. © 2023 Ford Motor Company.

Race-prepped for two years, Ford announced that the 2023 Bronco DR would be available for consumer purchase, turnkey-ready for brutal, high-speed desert runs in any off-road competition or purely for personal enjoyment. The DR features a Coyote V-8 5.0L (302 cu in), perhaps a tie-in to the exclusive 50 units to be built and to pay homage to Ford's racing heritage. © 2023 Ford Motor Company.

Having Hall as a last name may open a door or two, but in the competitive world of motorsports, that's all it will do. Proving you can live up to the legacy of Grandfather Rod Hall requires determination, skill, strength, and fearlessness. Shelby Hall embodies all of that and much more. She's got a winning record off-road, great marketing flair, and the kind of personality that will light up a room or a racetrack. There's a reason Shelby is in demand today and a big part of Ford's race program.

Growing up in a family that knew its way around a Bronco and a trail, hill, and mountain or two, Shelby started driving vehicles in the dirt as a child. Both "Papa" Rod and her father Josh ran an off-road driving school, and she was naturally along for the ride—as well as picking up bits and pieces of instruction here and there that stuck, well, like mud to a race truck.

Shelby's skills quickly grew—under the tutelage and ultimately, in partnership with Rod—and she became a professional driver, trainer, event facilitator, and spokesperson. She spent seven years as the administrator of the Off-Road Motorsports Hall of Fame. During that time, she also raced in off-road competitions with her grandfather. She was navigating during an initial run with him when he convinced her to move behind the wheel, "throwing me to the wolves and into the driver's seat," she noted. More recently, Shelby started her own business, Shelby Hall Off Road, hosting off-road events, conducting training workshops, and building a race UTV.

In 2019, Shelby joined the Bronco R race team for 6th gen's inaugural Baja 1000 race. She returned in 2020, and her reputation, skill, and experience led Ford to secure her talents as a Brand Ambassador, making appearances on behalf of the company as well as serving as a consultant for female-based driving programs. Over the past four years Shelby has competed in numerous endurance races and became the first driver to claim a victory in a Bronco 6th gen at the 2020 Rebelle Rally where she and Navigator Penny Dale traveled over 1,200 miles on a seven day trek. Shelby will return in 2023 with new teammate Rori Lewis.

Shelby recently signed with Ford's motorsports arm, Ford Performance, which will see her representing the Blue Oval while racing off-road and elsewhere. She stated on social media, "It's been a goal of mine to wheel more often this year," and in 2023 she has managed to participate in multiple desert events including the Mint 400, Rage at the River, NORRA, and the "romp from Ensenada to Cabo." Sounds like a very sensible plan. Rod would approve.

The incredible Bronco DR prototype gains some height and "stretches its legs" a bit as Ford engineers and professional drivers examine its capabilities as a race vehicle. © 2023 Ford Motor Company.

OPPOSITE: Preparing to tackle the Baja 1000 in the Bronco R, Shelby Hall stands by her steed proudly as one of the team's drivers. © 2023 Ford Motor Company and Shelby Hall.

A WORLDWIDE COMMITMENT

Ford takes its racing seriously around the world. It sees the value of motorsports as both a tool for research and development and marketing, in addition to the end-result of all promotional efforts, sales. Henry Ford saw racing as a way to prove the validity of his product—in this case, the Ford 999, his purpose-built race car—by having famed driver Barney Oldfield win against fellow competitor Alex Winton at Grosse Point in 1902. That enabled the determined automaker, due to the resulting publicity, to gain the financial backing he needed to establish the Ford Motor Company.

The manufacturer continues to see gains achieved through Ford Performance, its motorsports arm, and to that end, will expand its goals with a return to Formula One (F1) in partnership with Oracle Red Bull Racing as the power unit supplier to both Red Bull and Scuderia AlphaTauri teams starting in 2026. Ford last competed in F1 with Jaguar Racing in 2004 with drivers Mark Webber and Christian Klein and earned its 176th Grand Prix victory—its most recent—in 2003 from Giancarlo Fisichella in a Jordan Ford at the Brazilian Grand Prix.

"This is the start of a thrilling new chapter in Ford's motorsports story," declared Bill Ford, Executive Chair, reminiscing how his great-grandfather's racing efforts 120 years ago got the company going. "Ford, alongside World Champions Oracle Red Bull Racing, is returning to the pinnacle of the sport, bringing Ford's long tradition of innovation, sustainability and electrification to one of the world's most visible stages."

Ford continues its motorsports involvement in all facets of racing in all corners of the world, from grassroots to F1, highlighting endurance competitions and its vehicles' strength and durability, with Mustang, Puma, Ranger, Raptor, and Bronco. Within these multipronged and multiyear investments and programs, the company's engineers leverage Ford's efforts and outcomes to speed innovations, achieve real-world outcomes through the use of the latest technologies and software, and apply them to improve upon the vehicles the Blue Oval offers to its customers. Bronco and its siblings will only get better as a result.

And that's what competition—and winning—are all about.

Carry On

For some, the June date in 1996 was a sad one—for factory workers, Ford marketers, Bronco owners, and dealers. Some saw the glass as half empty, others as half full. Of course, another vehicle was filling the void at MAP. Expedition would occupy the marketers' efforts. Some Bronco owners dreamed their vehicles would increase in value since they would hopefully become rare. Dealers did too, but they were in the business of selling new cars and trucks and didn't have time to reflect on yesterday.

Still, it was the end of an era. The last Bronco built at that time was a 1996 XLT Sport in Oxford White Clearcoat. No news on where it went, though it was commemorated with an official plaque.

Yet, as discussed throughout this book, the idea—and the desire—of another Bronco emerging from a Ford factory never died. When the green light did come in 2015 from Ford leadership, no doubt there was a great deal of low-key cheering within corners of the Blue Oval who knew, replaced by respectful applause at the teaser announcement in 2017, and finally boisterous support at the 2020 launch party. No doubt a standing ovation was held upon the initial roll-out at the factory in early 2021 and start of production in June 2021, some 25 years later.

It had taken a while, but good things come to those who wait. For all of us who love Broncos, it's a happy time and one that hopefully has no ending.

This book, however, does. Thanks for taking a ride.

APPENDICES

Appendix 1: Bronco Clubs Across the Country

Here you'll find a multitude of Bronco clubs but certainly not all of them. New ones appear regularly, which is proof that Broncos have an incredible appeal and that the focal point continues to be about building connections and expressing one's passion for a lifestyle that incorporates their vehicle, whatever generation it may be.

Thank you to bronco6g.com/forum/clubs-groups for its help. To find a specific club's forum, Facebook page, or website, visit the link from Bronco6g.com and scroll through the more than five pages of clubs and groups listed. Several examples of specific links are listed below.

Adirondack Bronco Club *(Apparently, this is the winner for first listing.)* https://www.bronco6g.com/forum/groups/adirondack-bronco-club.141/

Alabama Bronco Club *(Alabama is usually first.)* https://www.facebook.com/groups/3519834701373445

Alberta Bronco Club (A.B.C.) *(Not going to be first.)* www.albertabroncoclub.ca (coming soon)

Anne Arundel County MD Broncos *(They tried hard to be first, but no posts yet.)* https://www.bronco6g.com/forum/groups/anne-arundel-county-md-broncos.157/

Arizona Bronco Club

Arizona Classic Bronco Club (ACB)

Arkansas Broncos

Bad Broncos *(fun group?)*

Badlands Club

Base Bronco Club

Bay Area Broncos

Bayou Bronco Club

BC (British Columbia) Broncos

Berta Broncs

Big Bend Club

Birmingham Bronco Society *(not a club, but a "society")*

Birmingham Broncos

Black Diamond *(not skiers)*

Blue Ridge Broncos

Bo Beuckman Buyers *(Buyers and reservation holders from Bo, unite!)*

Boise Bronco Owners

Born Wild & Wonderful Bronco Club (WV)

Brevard Broncos

Bronco Babes

Bronco Club of America *(grandaddy)* https://broncoclubofamerica.com

Bronco Club of Madison

Bronco Club of Northern Illinois

Bronco Club Orlando

Bronco Club UK/Europe (EUK) *(Let the Old Country have some fun too.)*

Bronco Club of Windsor and Essex County *(Across the river from Detroit.)*

Bronco Decatur, IL *(Let's keep it very specific.)*

Bronco Enthusiasts of Connecticut

Bronco Everglades

Bronco Gals

Bronco Ladies

Broncos of the Carolinas

Bronco Owners Association of Richmond (VA) *(BOAR is the club acronym; no reflection on the type of people or meetings they have.)*

BroncOhio *(cool spelling choice)*

BRONC (Bronco Riders of Nebraska Club)

BroncoII4×4.com

Broncos of Pennsylvania

Broncos of Santa Barbara

Broncos of the Bluegrass *(Kentuckians and anyone who likes Bluegrass)*

Broncos of the Carolinas

Bronco's of Richmond

Bronco Owners of North East States (BONES) https://www.facebook.com/groups/98225088885/

Broncos of the South

Bronco757 (Hampton Roads, Tidewater, Virgina)

Born Wild & Wonderful Bronco Club (WV) *(honorable mention for best name)*

Canadian Bronco Club *(eh!)*

Canuck Bronco Club *(eh, deux)*

Carolina Broncos

Central California Broncos

Central Florida Broncos

Central Ohio Broncos
    https://www.bronco6g.com/forum/groups/
    central-ohio-broncos.87/
Central TX (Texas) Bronco Club
Charlotte (NC) Broncos
Chesapeake Bronco Club
    https://www.bronco6g.com/forum/
    group-posts/7051/?page=1
Chicago Area Broncos
Cincinnati Bronco Club
Classic Broncos Hawaii *(shaka!)*
Colorado Bronco Club
Colorado Bronco Divas *(hmm . . .)*
Colorado Classic Broncos
Colorado Ford Bronco Club (CFBC)
'Cuse (for Syracuse) & Greater CNY Bronco Club
Custom Bronco Club
Daytona Beach Area Bronco Club
Deep South EB Club
DFW Bronco Club *(flight crews only?)*
Dirty Jerz *(for people of New Jersey, or in this case, Jerz)*
Early Bronco Entertainment
Early Bronco Registry (EBR)
Emerald Coast Bronco Club (NW Florida)
Exploring Western N.C.—Day Rides—Bronco
    Round-Up!
First Batch MY 22 (first production model year 2022)
First Edition Squad
5/2 Build Crew *(another very specific club of members who had
    a build date of the week of 5/2—any year?)*
Ford Bronco Raptor Owners
Ford Bronco Sport Owners of FLORIDA
Four Corners Bronco Club *(cool place to visit)*
4 × Ford Bronco Club
FullsizeBronco.com (FSB)
Gold Country Broncos
Golden Horseshoe Buckaroos (Southern Ontario, Canada)
Golden State Bronco Club
Georgia Bronco
Georgia Broncos Club
Granger Fan Club—Broncotown, USA
    (fans of Granger Ford too)
Gulf Coast Bronco Association LLC
HABO—Houston Area Bronco Owners
Hampon Roads Bronco Club
Hawaii Bronco Club *(another shaka!)*

Hillbilly Bronco Club *(no moonshine involved)*
Idaho Bronco
Indiana Bronco
Jersey/East Cost (sic) Bronco *(We think they mean
    East Coast.)*
Kansas Bronco Club
    https://www.bronco6g.com/forum/groups/
    kansas-bronco-club.99/
Kentuckiana Bronco Club *(51st state?)*
Kentucky Broncos *(see?)*
Las Vegas Broncos
Leonardtown Ford Reservation Holders
Lonestar Early Bronco Club (LEBC)
Long Island Broncos
Louisiana Bronco Crew
Manual Bronco Owners *(What's a manual?)*
Manitoba Bronco Club
Maryland Bronco Club
Michigan Bronco Club
    https://forums.michiganbroncos.club/index.php
Michigan 4×4
Mid Atlantic Bronco Club
Mid Atlantic Early Bronco *(MEB, much better than meh)*
Middle Georgia Broncos
Midwest Broncos Bros & Sis
Midwest Ford Bronco Club
Military Bronco's
Mississippi Bronco Club
Mississippi Broncos
Mullinax Kissimmee Res Holders
Mullinax West Palm Beach
Nebraska Bronco Club
N.E. Florida Bronco Club
Nevada Bronco Club
New England Broncos
    https://www.bronco6g.com/forum/groups/
    new-england-broncos.21/
New England Classic Broncos
    https://www.facebook.com/
    NewEnglandClassicBroncos/
NJ Bronco Club *(Maybe this Jersey group can help with
    spell check for the other Jersey clubs.)*
NOCO Bronco (Northern Colorado)
Nor Cal Broncos
North Carolina Broncos

Northeast Georgia Mountain Broncos

Northeast Ohio Bronco Owners

Northeast Trail Ride Connections

Florida Broncos

North Shore (New Brunswick, Canada) Broncos
  (no, not Hawaii)

North Texas Bronco Club
  https://www.northtexasbroncos.org/
  (some are registered as nonprofits)

Northwest Florida Early Bronco Club

North West Classic Broncos

Northwest Territories ARCTIC Broncos *(Courtney Barber, next time, say hello!)*

Nova Scotia Bronco Club

NWA Broncos (Northwest Arkansas)

Ohio Bronco Club

Oklahoma Bronco Club

Oklahoma Classic Broncos

Ontario (Canada) Bronco Club

Oregon Bronco Clu×b

Outer Banks Club

Outer Banks Beach Club *(twice the fun)*
  https://www.bronco6g.com/forum/groups/
  outer-banks-beach-club.7/

Oxford White Bronco Club

Palm Coast Florida Broncos

Pittsburgh/Tri-State Bronco Club
  (Another attempt at three states; see below.)

Puerto Rico Bronco Club
  (maybe 52nd state after Kentuckiana)

Puget Sound Bronco Brigade

Queen City (Cincinnati) Bronco Club

Raptors R US *(apparently, just Bronco Raptors)*

Sac (Sacramento) Bronco Club

Sacramento Bronco Club *(Two of them in the same city?)*

S. TX Bronco Club

San Antonio Bronco Club

Scottsdale/Phoenix Bronco Club

SE Michigan Mud, Sweat, & Gears 4×4 Club

775 BC *(Winner for best club name—Bronco enthusiasts in Northern Nevada.)*

SHOW ME MO (Missouri) BRONCO

SoCal Broncos

Southern IL Bronco Club

South Florida (FTL/Miami to Key West) Bronco Club

Southwest Florida Broncos

Southeast Trans-America Trail

Stamford Ford, CT *(Dedicated to anyone who has ordered their Bronco from this dealership—nice!)*

Tallahassee Bronco Club

Tampa Bay Area Bronco Club

Tennessee Bronco Club

Tidewater Bronco Club *(Chincoteague to Elizabeth City)*

The Bronco Nation
  https://thebronconation.com/

The Broncos of Virginia(s) *(Apparently, both Virginia and West Virginia.)*
  https://www.bronco6g.com/forum/groups/
  the-broncos-of-virginia-s.64/

The Montana Bronco League

The PNW (Pacific Northwest) Bronco Group

Tidewater Broncos (lots of Virginia clubs)

Treasure Coast and Palm Beaches

Tri-State Broncos *(New York, New Jersey, Connecticut, and Pennsylvania—really quad-state, but we didn't start the club.)*

Tri-State MD VA DC Bronco
  (Technically, DC is NOT a state.)

2-Big Broncos 4×4 Club

Utah Bronco Club

Utah Broncos

Utah Broncos Owners Unite!

Vegas Broncos

Virginia Broncos

Western Basqlands (Badlands Sasquatch owners)

West Tennessee Bronco 6g

West Texas Broncos

Wichita Bronco Club
  (No, it's not pronounced Wi-chee-tah.)

Wildtrak Owners Club

Windrock Broncos

WNY (Western New York) Bronco Group

Wisconsin Bronco Club

Wyoming Bronco Club *(last, but surely not the least)*
  https://www.bronco6g.com/forum/groups/
  wyoming-bronco-club.160/

The comments in parentheses are from the author
of this book and intended for levity and not to offend.

Appendix II: Selected Bronco Events Across the Nation:

Bronco Annual Event Schedule (check each website for future event dates)
For a comprehensive list of all Bronco activities, check out https://thebronconation.com/events/
or https://tomsoffroad.com/pages/upcoming-events/

DATE	HOST	STATE	URL
First week of March	United By Bronco Off Road Event	Hurricane, UT	unitedbybronco.com
March 15-19	Bronco Stampede	Rock Springs, AZ	Tomsoffroad.com/pages/upcoming-events/
April 13-16	Lonestar Early Bronco Texas Round Up	Mason/Katemcy, TX	lebc.clubexpress.com
April 17-20, 2024	Bronco Super Celebration East	Townsend, TN	https://www.broncodriver.com/index.php/super-celebration/
May 2-6	Moab Bronco Safari	Moab, UT	broncosafari.com
Mid-May	9th Annual Wild Horses Round Up (in 2024)	Lodi, CA	wildhorses4×4.com
May 6	25th Annual Bronco Swap Meet	Hudsonville, MI	broncoswapmeet.com
May 12-13	May It Forward	Gardner, MA	thebronconation.com
May 26-27	BroncoFest '23	Wytheville, VA	swvabroncofest.com
June 2-4	Carlisle All Ford Nationals	Carlisle, PA	earlybroncos.com
June 21-24	Mid-Atlantic Early Broncos (MEB) Round Up (26th Annual; 2024)	Pine Grove, PA	earlybroncos.com
July 12-15	Bronco Super Celebration Wisconsin	Wisconsin Dells, WI	https://www.broncodriver.com/index.php/super-celebration-wisconsin/
July 20-22	Tom's Bronco Parts Rock & Roll	Medford, OR	tomsoffroad.com
August 4-5	Bronco Take Over	Mears, MI	broncotakeover.net
August 16-19	Northwest Bronco Round Up	Florence, OR	northwestbroncoroundup.com
August 18-19	Woodward Dream Cruise	Pleasant Ridge, MI	thebronconation.com
September 5-9	2023 Bronco Super Celebration West	Buena Vista, CO	https://www.broncodriver.com/index.php/super-celebration-west/
September 20-23	OK Classic Bronco Round Up	Hot Springs, AR	okclassicbroncos.net
October 4-7	2023 Bronco Super Celebration at Carson City, NV	Carson City, NV	https://www.broncodriver.com/index.php/nevada-event/
October 5-7	Broncos at the Beach	Myrtle Beach, SC	earlybronco.com
October 12-15	Wheeling for a Cause	Mason/Katemcy, TX	wheelingforacause.org
October 20-21	Duffs, Diners and Drivers @ James Duff, Inc.	Knoxville, TN	Dufftuff.com
November 3-5	Bronco Daze Casual (new name)	Borrego Springs, CA	earlybronco.com
December 8-10	2023 Arizona Bronco Round-Up	Wickenburg, AZ	thebroncoroundup.com

Appendix III: References & Websites

Bronco- or Ford-related websites:

https://aikinsaviationart.com/contact.html

https://bajabronco.com

Bring Back Bronco, Episode 1
  https://www.youtube.com/
  watch?v=NYWbsrIL4mc

Broncos in TV and Film
  https://www.imcdb.org/vehicles.php?results-
  Style=asImages&sortBy=4&make=Ford&mod-
  el=Bronco&modelMatch=1&modelInclMo
  del=on&page=18

http://www.bronco.com/cms/early_bronco_history

https://www.broncodriver.com

https://www.broncoii4×4.com/forum/

The Legend Returns
  https://thebronconation.com/
  the-legend-returns-a-bronco-nation-film/

https://thebronconation.com/
  join-a-bronco-club-near-you/

https://broncooffroadeo.com/locations

https://www.broncoraptor.com/

https://www.bronco6g.com/forum/

https://www.broncosportforum.com/forum/
  threads/retro-70s-freewheelin-graphics-on-
  bronco-sport-badlands.7880/

https://ford.com

https://www.ford.com/bronco-wild-fund/

https://fordauthority.com/category/ford/bronco/

https://www.fullsizebronco.com/

https://www.gatewaybronco.com/ev-bronco/

https://www.hennesseyperformance.com/

https://icon4×4.com/about-icon

https://www.instagram.com/auto.nation.
  illustration/

https://www.joecottonford.com/
  history-of-the-ford-bronco/

John Bronco Rides Again—
  https://johnbronco.com/

https://www.maxliderbros.com/

https://www.ridefox.com/subhome.php?m=truck

https://www.thehenryford.org/

https://www.facebook.com/theshelbyhall/

https://trucktalkmedia.com/bronco-talk

https://www.unitedbybronco.com/

https://www.wendlefordsales.com/blogs/3550/
  easter-egg-hunt-in-the-ford-bronco/

Printed books and selected online resources

1984-1990 Ford Bronco II Specs & Facts. Bronco Corral. (2020, November 24).
  https://www.broncocorral.com/tech_library/ford-bronco-ii-facts/

Elkin, J. (2022). *Bronco racing: Ford's legendary 4 by 4 in off-road
  competition*. CarTech, Inc. Grisham, L. (2014, June 17). *For Ford's
  Bronco, O. J. Simpson chase may have helped sales*. USA Today.
  https://www.usatoday.com/story/news/nation-now/2014/06/17/
  the-bronco-brand-after-oj/10257945/

Herbez, V. (2020, December 1). *Ford Bronco History (1965-1996)*.
  Bronco Bastards. https://broncobastards.com/blogs/history-ford-bronco/
  ford-bronco-history-1965-1996

Jablansky, J. (2018, January 5). *The complete history of the Ford Bronco*.
  The complete history of the Ford Bronco Men's Journal. https://www.
  mensjournal.com/gear/the-complete-history-of-the-ford-bronco-w460265

Jurnecka, R. (2020, July 21). *Ford Bronco: history, specifications, models,
  buying tips, news.*

MotorTrend. https://www.motortrend.com/features/
  ford-bronco-buying-guide-specs- details/

Mellor, D. (2022, March 6). *The 1966 Ford Bronco launch... and what's
  coming next*. Unsealed 4 by 4. https://unsealed4×4.com.au/
  the-1966-ford-bronco-launch-and-whats-coming-next/

Palmer, Z. (2020, July 14). *2021 Ford Bronco trim level breakdown:
  features of Big Bend, Black Diamond, Outer Banks, Wildtrak and Badlands*.
  Autoblog. https://www.autoblog.com/2020/07/14/
  2021-ford-bronco-trim-level-breakdown/

Remembering Donald N. Frey—designer of the Mustang. UMich
  MSE. (2010, March 6). https://mse.engin.umich.edu/about/news/
  remembering-donald-n-frey-designer-of-the- mustang

Shaw, K. V. (2022, September 23). *How Shelby Hall, granddaughter of
  famed off-roader Rod Hall, is creating her own legacy*. The Drive.
  https://www.thedrive.com/news/how-shelby- hall-granddaughter-
  of-famed-off-roader-rod-hall-is-creating-her-own-legacy

Stone, M. (2021, May 28). *Bill Stroppe's Baja Broncos; off-roading legend,
  on- and off-road.*

Matt Stone Cars. https://mattstonecars.com/
  bill-stroppes-baja-broncos-off-roading-legend-on-and-off-road/

Udy, J. (2019, June 18). *Three decades of the Ford Explorer: A look back
  at the SUV's history*. MotorTrend. https://www.motortrend.com/
  features/25-years-of-the-ford-explorer-a-look-back-at-this-suvs-history/

Wasef, B. (2021, February 23). *How McKinley Thompson, the first major
  black car designer, led Ford into the future*. Robb Report.
  https://robbreport.com/motors/cars/mckinley-thompson-first-major-
  black-automotive-designer-ford-1234595635/

Zuercher, T. (2019). *Ford Bronco: A history of Ford's legendary 4 by 4*. CarTech Inc.

ACKNOWLEDGMENTS

First off let me thank Zack Miller, Motorbooks' Group Publisher, for allowing me the chance to write this Bronco history. I have had the honor of working with Motorbooks on two previous books, both on the Nissan Z and each containing original material, and they were great experiences.

When the new Bronco was announced, I pitched Zack to write its story. I am very glad the project came to fruition. It was a pleasure to work with Quarto Publishing Group, especially Project Manager Brooke Pelletier and Senior Marketing Manager Steve Roth.

Second I have to thank my contacts at Ford, particularly Ted Ryan, Archives and Heritage Manager, for all his help and the great images and info he, Jamie Myler, Research Archivist, and Ciera Castell, Collections Archivist, sent my way. Thank you to Mike Levine in Public Relations, who came to the dance late, but was extremely helpful in my quest for info. I also want to extend my appreciation to Mark Grueber, Bronco Marketing Manager, for his time in answering my many questions.

I'd like to give a shout-out to publicist, storyteller, and incredible car enthusiast Jiyan Cadiz, who left Ford in the middle of my work to pursue other interests but was nonetheless extremely helpful up to his last minute—and beyond.

Ford deserves credit for its openness in providing virtually anyone with the opportunity to seek out images to download through its Heritage Vault, a wonderful new database that makes finding something that feeds your passion about Ford easily available. A tip of the cap for making this happen, Blue Oval staffers.

"Ford created the Heritage Vault (https://fordheritage vault.com) for fans, journalists and car enthusiasts, making it easy to discover the company's rich heritage from anywhere in the world. The artifacts are downloadable for personal use, free of charge, for the first time as part of the American auto industry's most comprehensive online database."

My deepest appreciation to Shelby Hall, who kindly agreed to my request to write the foreword to this book. Shelby possesses the DNA of a legend (off-road racing legend Rod Hall is her grandfather) and has become quite the Bronco ambassador. Posting her own race wins, she has quickly shown that the apple doesn't fall far from the tree. I know she'll continue to do great things throughout her life, and I'm honored to have had her involved with this project.

Additionally I must thank all the folks who were kind enough to share a story or answer a question about how Bronco (or their Bronco) had made an impact on their lives or allowed them to start a related business. Thanks, as well as the folks on the forums, Facebook pages, Bring a Trailer listings, and others who kindly shared their photos and a little information for inclusion in this book. Your involvement was key to making the content herein and discussions all about the lifestyle and connections that Bronco helps generate. Jordan Parker of Bronco Nation; Dave Landwehr of Bronco6g.com; Jack Vest, President, Alabama Bronco Society; as well as Jake Gertsch of Montana Broncos; Ronnie and Autumn Welch of Truck Talk Media and Bronco Talk; and Chau Nguyen of Vintage Broncos are among those who deserve a special shout-out. I'm very grateful for your time and support, folks.

Finally to my amazing wife who endured many dull weekends when I sat in my mancave writing and editing this manuscript, making sure I poured my soul and corresponding zeal into it. Thanks for understanding me. I promise once this is done, we'll do more outdoor stuff together—maybe in a Bronco, okay? A trip to the Greek Islands, you say? Yes, dear.

INDEX

Page numbers in *italic* type indicate images